Take Note

- Descriptions in this manual are bas… Samsung Galaxy A12.
- All the instructions in this book is … of Samsung Galaxy A12, and not advisabl… out on any other Samsung mobile device.
- Although I made sure that all information provided in this guide are correct, I will welcome your suggestions if you find out that any information provided in this guide is inadequate or you find a better way of doing some of the actions mentioned in this guide. All suggestions should be forwarded to SodiqTade@gmail.com.

Copyright and Trademarks

About This Guide

Finally, a detailed guide on Samsung Galaxy A12 here. This guide is the best for this new mobile device. Spiced with some cool features, tips and tricks with the recent advanced features.

This is a very simple and comprehensive guide, well detailed guide, useful for older adults, experts, newbies and iPhone switchers.

This guide contains a lot of information on the Samsung A series.

It is a step to step guide with screenshots, suggestions and notes. This guide is particularly useful for older adults, newbies and seniors; however, I believe that Pro and experts will find some benefits reading this guide.

Have a wonderful time as you read this detailed guide.

Your gift will be a complete one if you gift this guide together with the new Samsung Galaxy A12. Smiles.

Table of content

Turning Your Phone on/off

Just like many other smartphones, turning on your Samsung Galaxy A12 is as simple as ABC. To turn on your phone, press and hold the Power Key until you notice a small vibration. If you are turning on your phone for the first time, please carefully follow the on-screen instructions to set it up.

Note; It is advisable that you insert the SIM card before switching on your phone. To learn more about inserting the SIM card.

To turn off your phone, press and hold the power key and select **Power off.** Select **Power Off** again to confirm.

AS shown below

Quick way to switch of your device

- Swipe down from the top of the screen and select the power icon.

As shown below

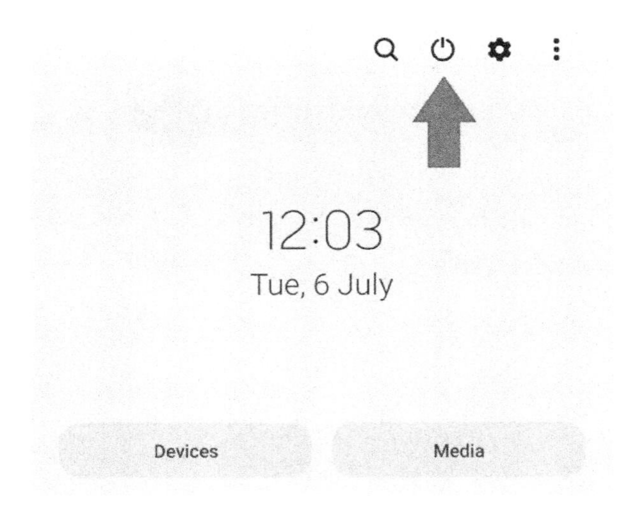

- Select the **Power Off** again to confirm

Please do not vex if you find it unnecessary reading about how to on/off your device. I have included it in this guide, in case there may be someone reading this guide who is a complete novice and knows close to nothing on smartphones.

Tips:

- During the setup, you may skip any process by tapping **SKIP** located at the bottom of the screen. Usually, you will have the option to perform this process in the future by going to your phone settings
- Because of software updates, it is likely that your device will consume a large amount of data during the setup, it is advisable that you connect to a wireless network if you can. Using a mobile network during the setup may be expensive.
- You will probably notice that your phone screen locks within few seconds after you finish using it. To allow your phone to stay longer before it locks, change the screen timeout setting. To get this:
  - Swipe down from the top of the screen and select Settings icon .
  - Tap **Display**.
- Scroll down and tap on **Screen Timeout**. Then choose any of the options.

As shown below

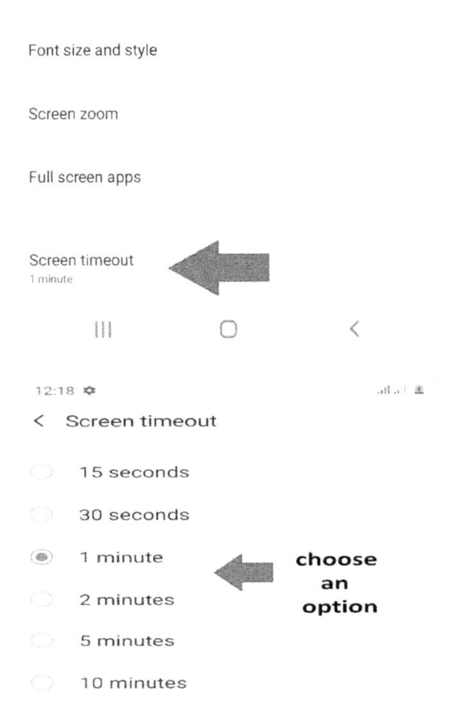

Note: Selecting longer time may make your battery discharge faster. Also it is not really necessary you charge your Samsung Galaxy A12 before the first use. And you can also charge it if there is a way to charge it before the first use.

It appeared that the new lithium batteries used in smartphones don't really need to be charged before first use.

Swiping the Screen Properly

You will need to interact with the screen of your phone by swiping it with your finger time to time. If you don't swipe it properly, you may not get the expected result.

Access the notification menu/quick settings

To access the notification menu, swipe from the top of the screen. Please make sure you are starting from the top of the screen to get the expected result.

Accessing the app screen

To access the applications screen, swipe up from the lower part of the screen. See the direction of the arrow below.

As shown below

Settings Tab

Settings tab is one the most used section on every device. And I will be referring to this section a lot.

The settings tab has many subsections, so it is advisable that you use the **Search** menu to quickly find and navigate what you are looking for. To use the setting's search feature, swipe down from the top of the screen and tap the settings icon.

As shown below

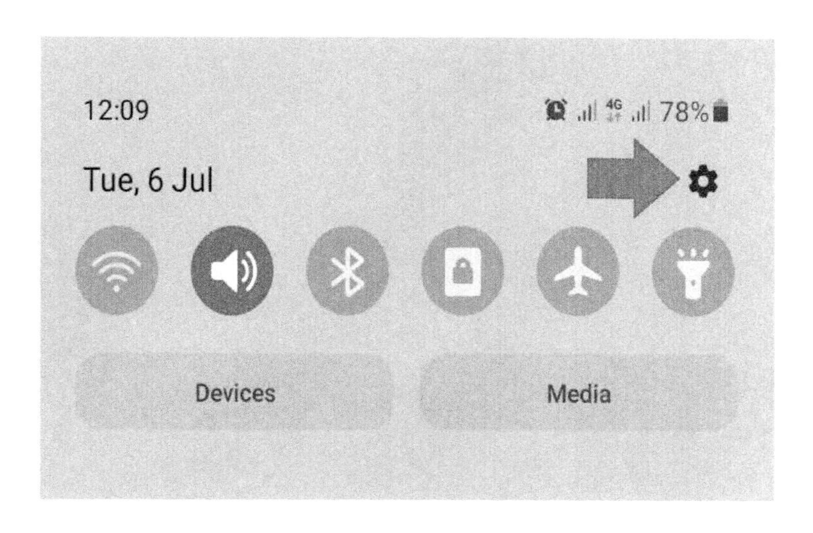

Then tap on the search icon and type a keyword corresponding to the settings you are looking for. For example, if you are looking for settings relating to data usage, just type **Data** in the search bar.

As shown below

12:24

Settings

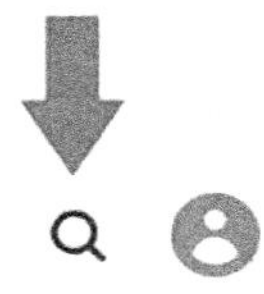

Connections

Wi-Fi, Bluetooth, Flight mode, Data usage

Sounds and vibration

Sound mode, Ringtone, Volume

Notifications

App notifications, Status bar, Do not disturb

Charge Your Device

It is cool to charge your Samsung Galaxy 12 anytime you are not using it, just to ensure you always have enough power on it. in case you want to play games, watch videos and do a lot of stuff on it later. One of the best time to charge your phone is when you are not using it. Maybe when you are taking a nap, in showers or busy with cooking.

To charge your Samsung Galaxy A12:

1. When you unbox, you will notice that the power cord consists of two parts (the USB cable and the power adaptor), connect these two parts together.
2. Connect the end of the USB cable to the charging port of your device, making sure that both the charging cable and the charging port on your device are well fixed.
3. Plug the power adapter to an electrical outlet. When your phone is charging, a charging icon will appear at the top of the screen. When your phone is fully charged, the battery icon will appear full.
4. Unplug when fully charged.

Note: After charging, you may need to apply a small force to remove the USB cord from the phone.

Fast Charging Feature

Your device is built with a battery charging technology that charges the battery faster by increasing the charging power. This feature can allow you to charge your Samsung Galaxy 12 up to 40% in about 35 minutes.

Note: Using your phone while charging may affect the time your phone is going to take to complete a charging cycle.

Inserting and Managing SD Card/SIM Card

Both Samsung Galaxy A12 support the use of external memory card.

To insert memory card or SIM card

1. Locate the SIM and SD Card tray right edge of the device and gently insert the eject tool/pin included with your phone into the eject hole (located at the right edge) and then push until the tray pops out.
   Note: You may need to apply small force before the tray pops out.
2. Use your hand to pull out the tray gently from the tray slot and place the SIM card on its tray and the SD card on its tray. Please make sure the gold contacts on the SIM and the SD card are facing down.
3. Slide the card tray back into the slot.

Please note that if the SIM card or the memory card is not inserted properly, your phone may not recognize it. Make sure you insert the SIM card and SD card properly.

Touch screen basics

Your phone's touch screen allows you to easily perform functions and navigate through your device.

Notes:

- Do not press the touch screen with your fingertips, or use sharp tools on the touch screen.

- When the touch screen is wet, endeavor to clean it with a dry towel before using it.

These will help protect your touch screen from malfunctioning.

Tips on how to use your touch screen:

Tap: Touch once with your finger to select or launch a menu, application or option.

Tap and hold: Tap an item and hold it for more than a second to open a list of options.

Tap and drag: Tap and drag with your finger, to move an item to a different location in the application grid/list.

How to Lock or Unlock the touch screen

When you do not use the device for a specified period, your device turns off the touch screen and automatically locks the touch screen so as to prevent any unwanted device operations and also save battery.

However, to manually lock the touch screen, press the power key. To unlock, turn on the screen by pressing the power key or double tap on the screen, and then swipe in any direction. If you have already set a lock screen password, you will be asked to enter the password instead of accessing your home page directly.

Rotating the touch screen

You may activate or deactivate screen rotation. To quickly disable or enable screen rotation, swipe down from the top of the screen and tap ____.

As shown below

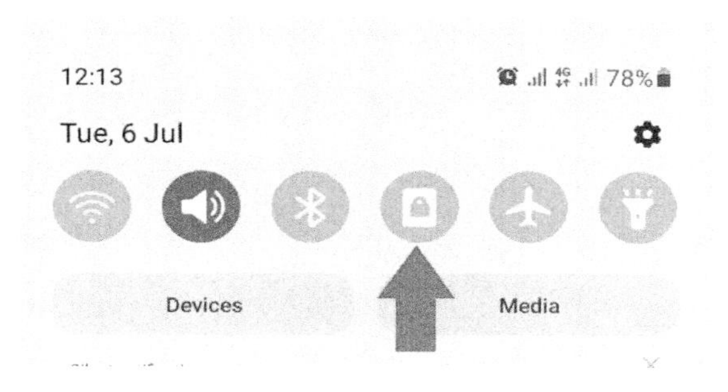

Navigating your device

Back Button

There is a dedicated back button (as shown below) to view the previous page or go back to a previous menu. Back button can also be used to close a dialog box, menu, or keyboard.

Moreover, you can use the dedicated back button on your Samsung device to get out of any page when you are done with the page.

Using the In-APP Back Button

There are some apps that give you the opportunity to go back to a previous screen using the in-app back button < . When available, this button can be found at the upper left part of the screen.

As shown below

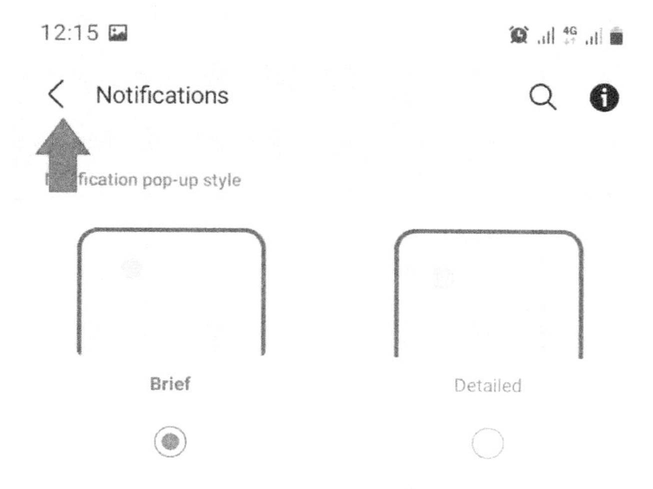

The Menu icon

The menu icon is the three dots icon ⋮ that usually appears at the top of the screen when you open an app.

As shown below

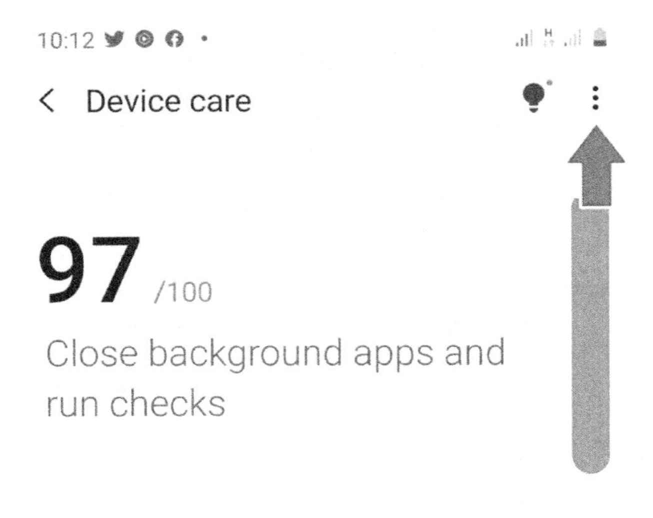

Customize the Home Screen

This allows you to customize your home screen to your taste.

To customize the home screen to your taste:

- Tap and hold an empty space on the home screen. (as shown below)

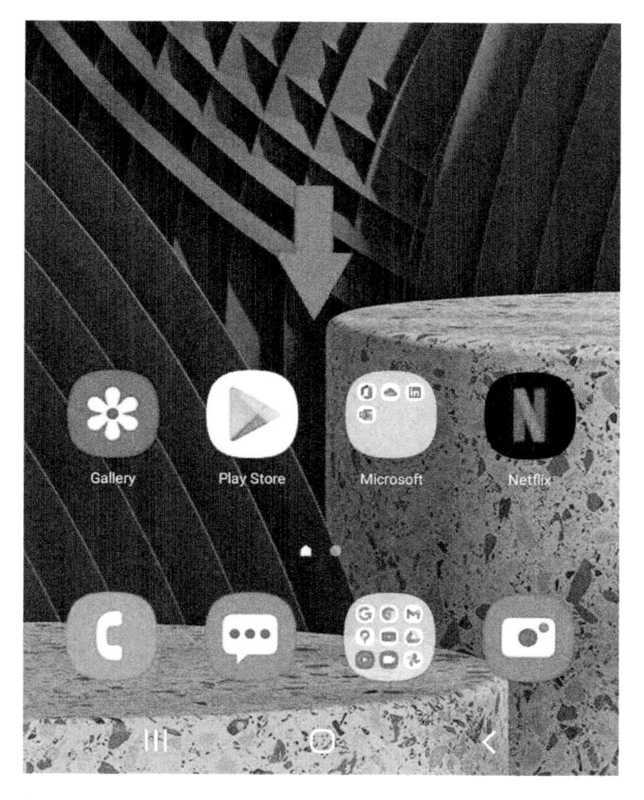

- To go to home screen from any screen, press the home button.

- Here you can perform some actions:
    - **Add a screen**: To do this, swipe left until you see the plus/Add **(+)** icon.

As shown below

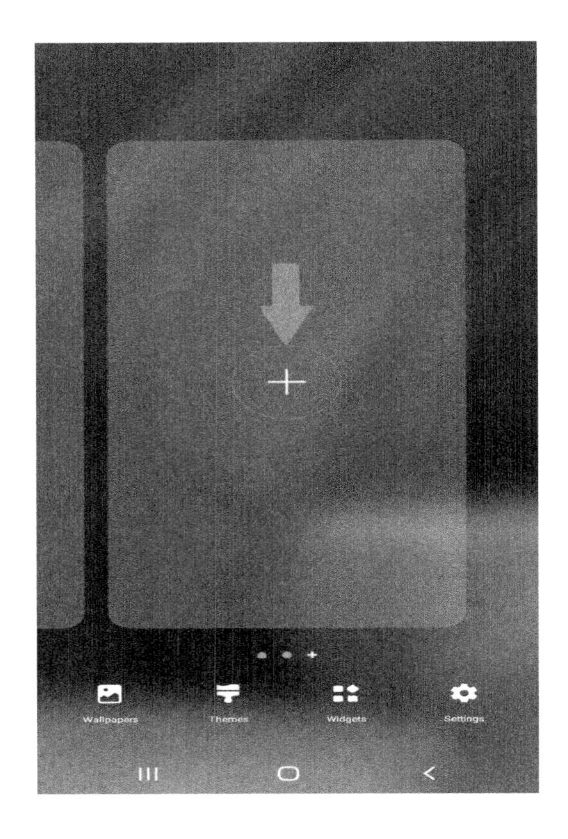

- **Remove a screen**: To do this, tap on the delete icon. As shown below

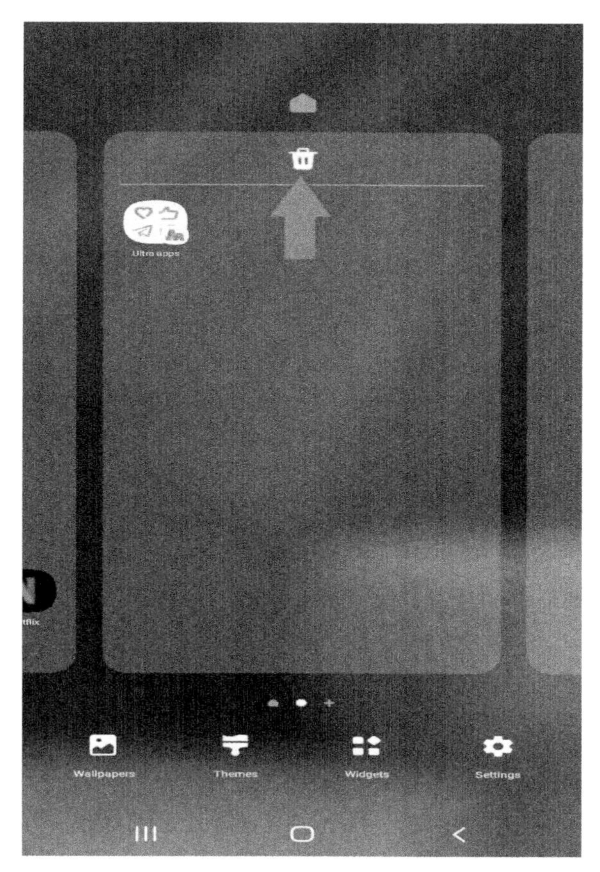

- **Change the order of the home screens:** To do this, tap and hold a screen and drag it to the either left or right of the screen until the page turns. Lift your finger when done.

Add/Remove apps to the home screen

You can add apps/items to the home screen so that you can easily access them anytime you need them.

To do this:

- Access the app screen by swiping up the screen while on the home screen.
- While in app screen, tap and hold an app icon, and then select **Add to Home**.

As shown below

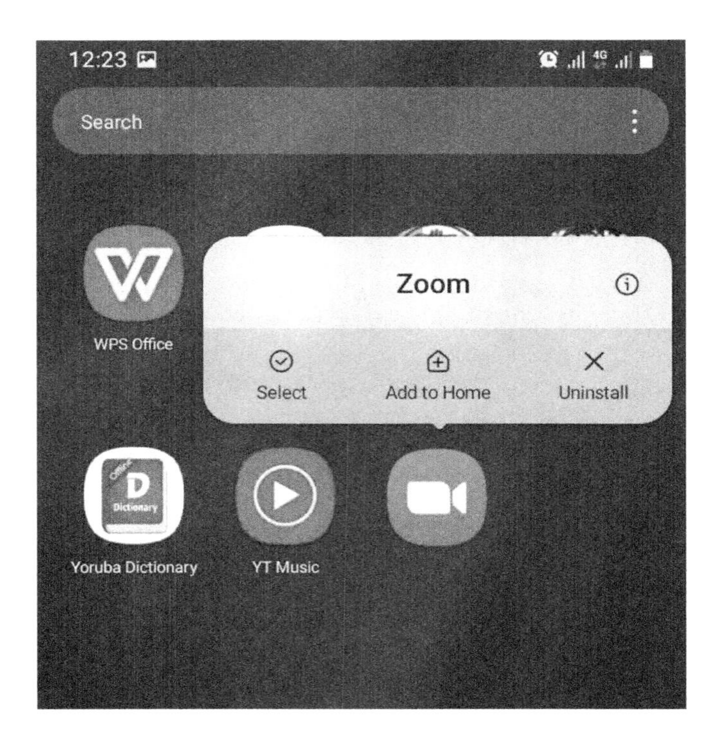

- To remove an app shortcut from the home screen, tap and hold the app icon you want to remove and then select **Remove from Home**.

As shown below

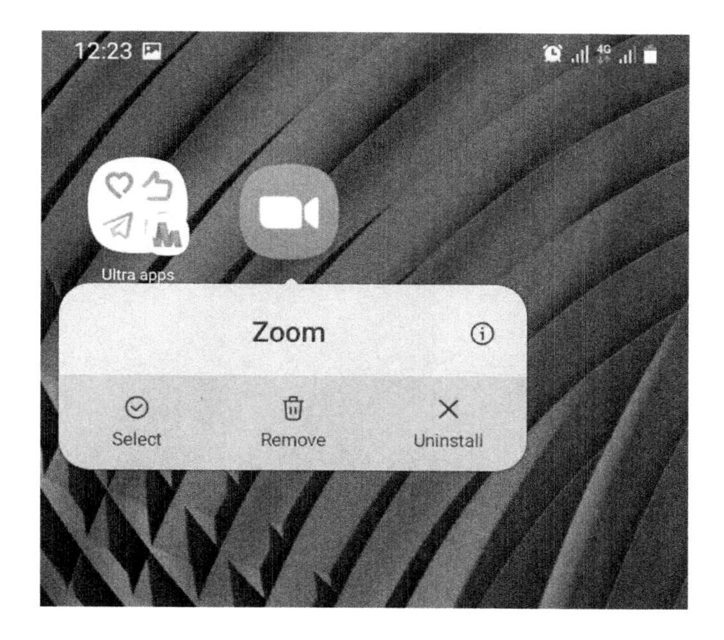

Note: Removing an app icon from the home screen does not uninstall the app. The removed app will still be in apps section.

Home Screen Layout:

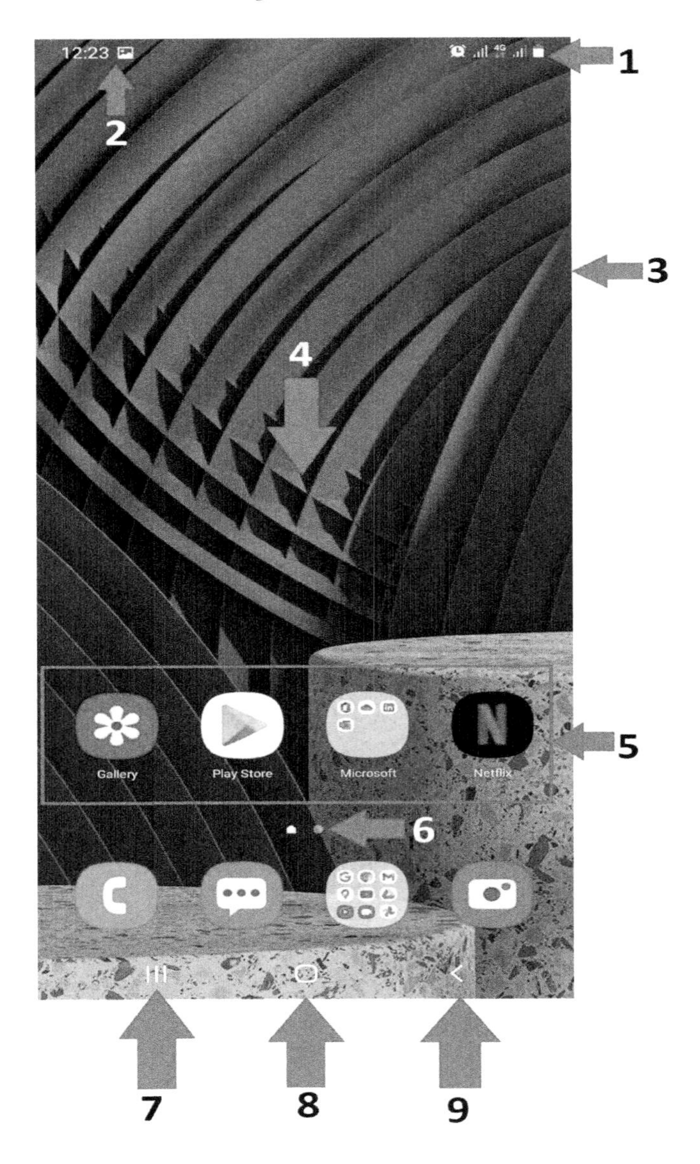

Number	Function
1.	Battery icon and Sim notification icon.
2.	Notification icon/ Status icons
3.	**Edge Handle**: Swipe left on this handle to display edge icons. You will see this only if you have enabled the feature on your device.
4.	Weather widget. It will only be visible if you have added it on your home screen.
5.	**Apps Shortcuts**: Launches your favorite application.
6.	**Home Screen Indicator:** This indicates which home screen is currently visible**.**
7.	**Recent App button**
8.	**Home Button**
9.	**Back Button**

To create a folder on the Home Screen:

From the home screen or application screen, tap and hold to select the apps you want to add in the folder. Then tap on **create folder** to create a folder. As shown below

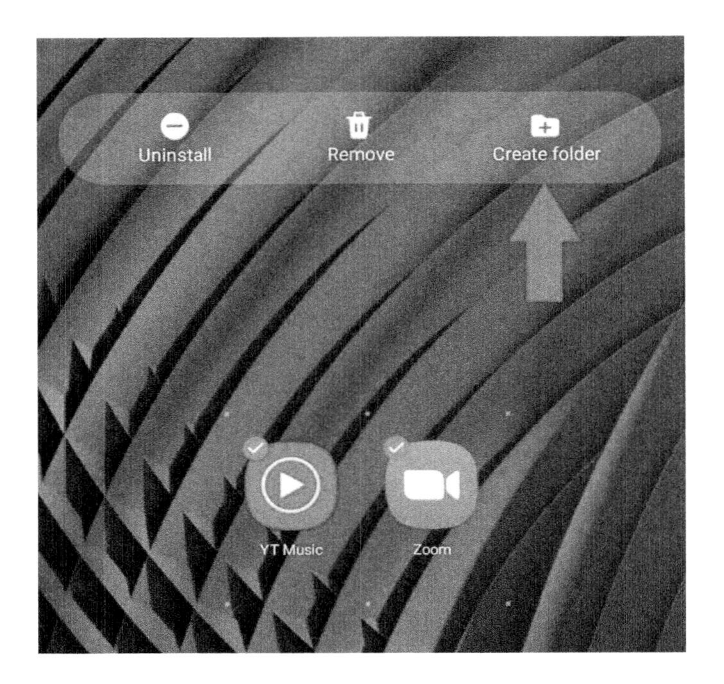

1. Tap **Enter folder name** and enter a folder name.
2. To change the color of the folder, tap on the tiny circle and select a color (as shown below).

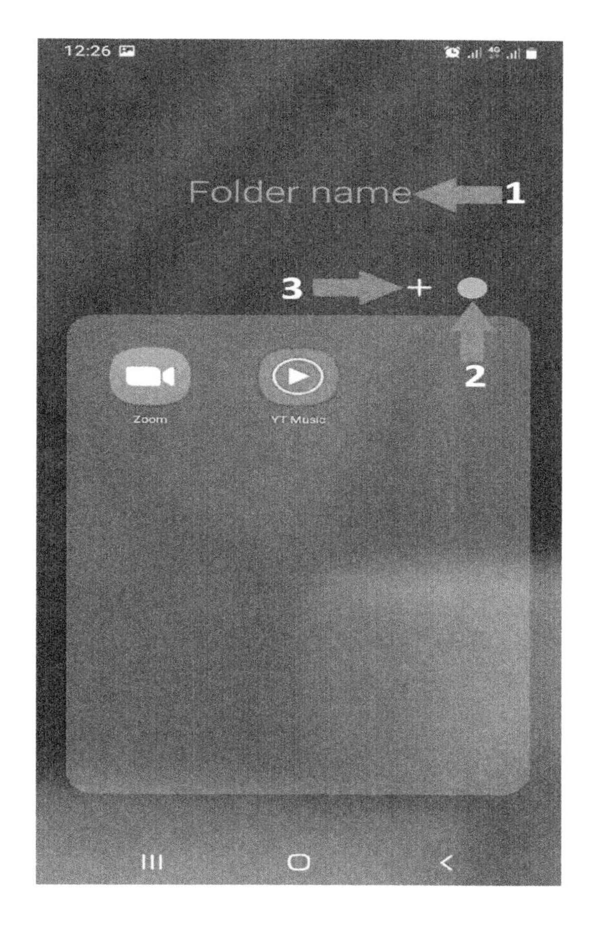

3. To add another app to the folder, tap the + icon located at the top right corner.
4. When you are done customizing a folder, tap anywhere outside of the folder or tap the **Done**.
5. To remove an app from a folder, tap the folder, and then tap and hold the app you want to remove and drag it out of the folder.

Accessing and Managing Applications

To open an app:

1. From the home screen; swipe up from the bottom of the screen to access application screen.
2. Tap on the app of your choice.
3. To go back to the app grid screen, press the back button.

As shown below

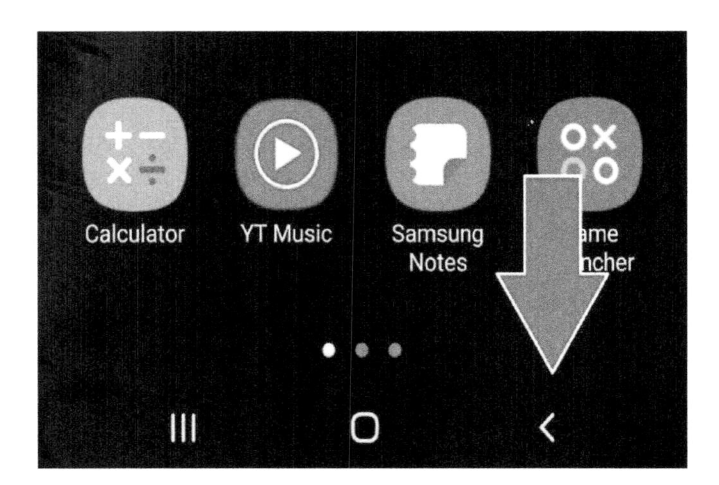

Accessing Recently Opened/Running Applications

- Tap on the recent button to show the recent app window. This contains the list of all opened/running apps.

As shown below

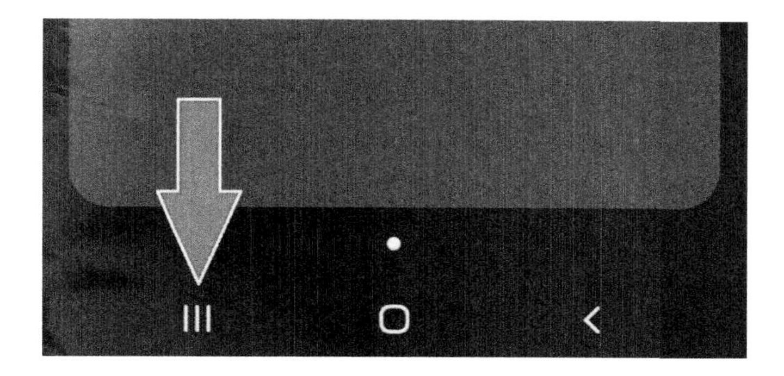

- Tap on the app to launch it, or swipe up to cancel/remove the app. To close all opened apps, tap **CLOSE ALL** located at the bottom of the screen.

Note: You can lock an app so that it does not get closed when you tap **CLOSE ALL.**

To do this:

To lock app

- Tap the recent button and select the app menu icon located at the top of the screen (as shown below). Then select **Lock apps.**

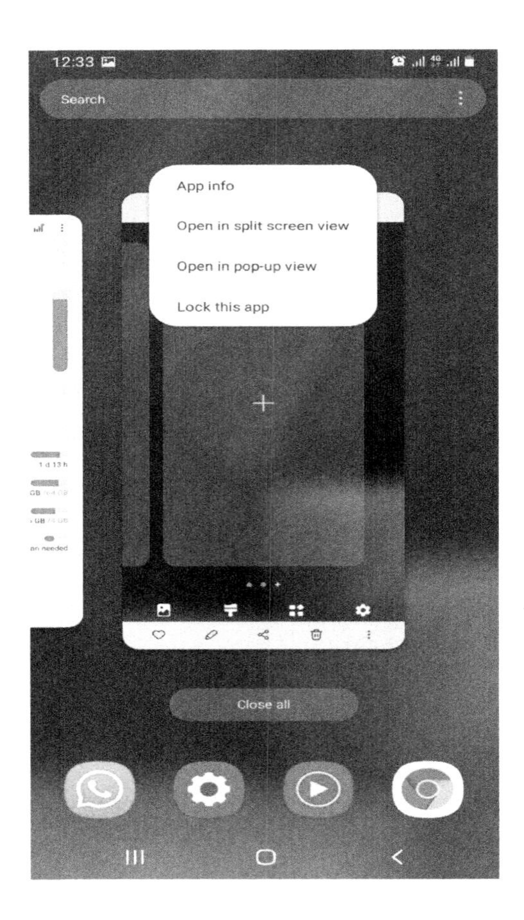

To Unlock App

- Tap the recent button and select the app menu icon located at the top of the screen. Then select **Unlock this app**. Or you can easily tap on the padlock icon to unlock such app.

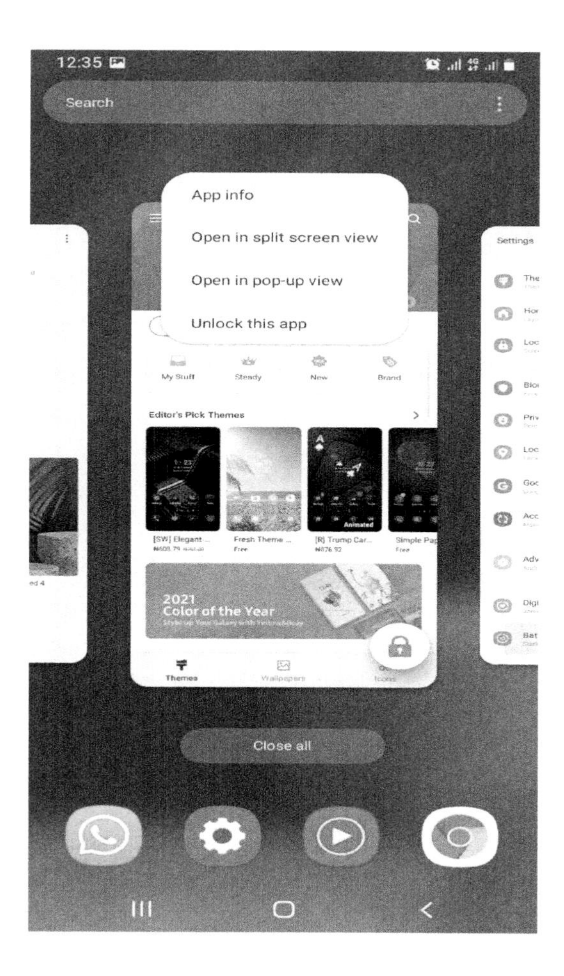

Show the apps button on the home screen.

To do this:

1. While on home screen, swipe up the screen to access the application screen.

2. Tap the menu icon at the top right corner and select **Home screen Settings** (as shown below).

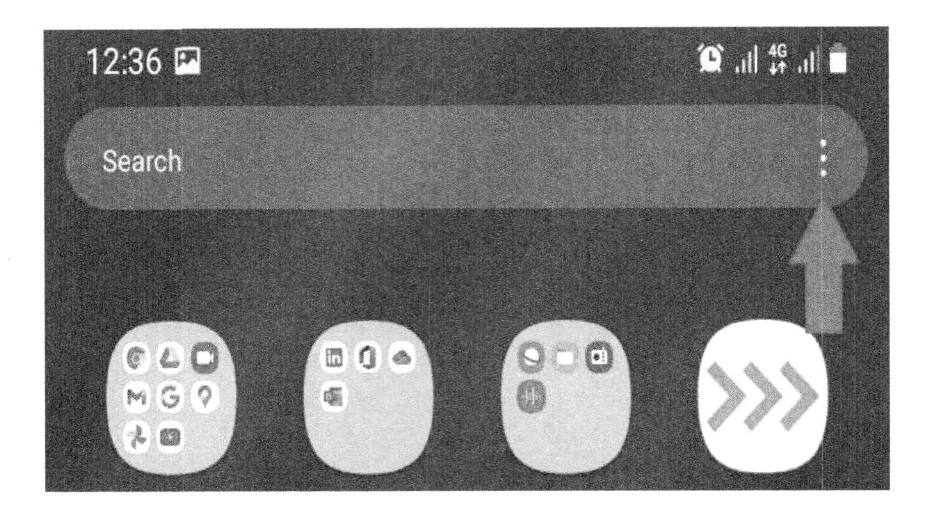

3. Toggle to enable Apps button.

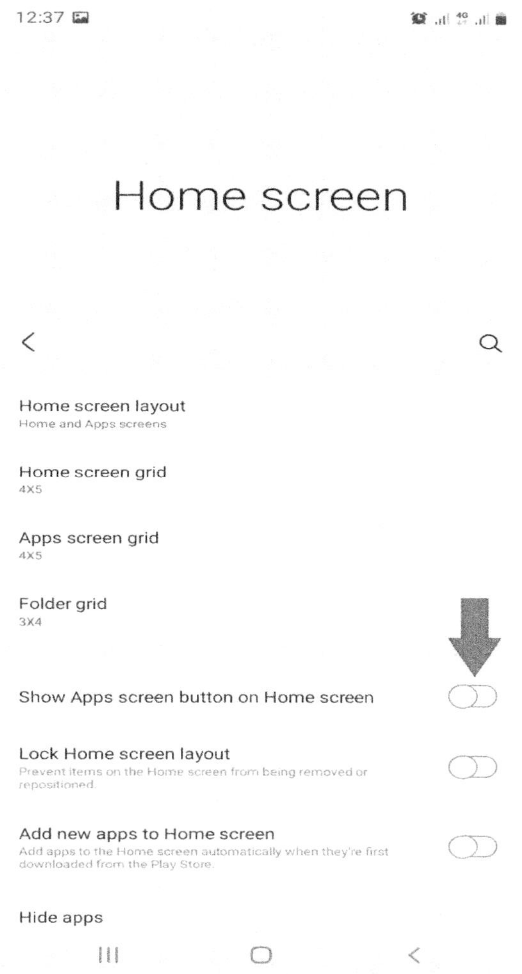

12:37

Home screen

Home screen layout
Home and Apps screens

Home screen grid
4X5

Apps screen grid
4X5

Folder grid
3X4

Show Apps screen button on Home screen

Lock Home screen layout
Prevent items on the Home screen from being removed or repositioned.

Add new apps to Home screen
Add apps to the Home screen automatically when they're first downloaded from the Play Store.

Hide apps

How to hide applications

If you don't want your friends/kids to access sensitive apps on your phone, you can hide them. For example, if you don't want your children to access your messenger app, you can hide them.

To hide apps:

1. While on home screen, swipe up the screen to access the application screen.

2. Tap the menu icon ⋮ and select **Settings**.

As shown below

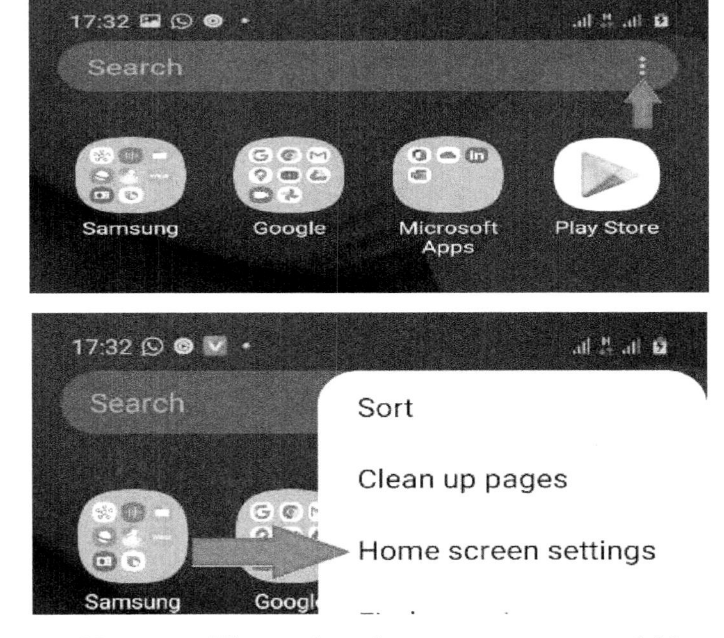

3. Tap **Hide apps.** Then select the apps you want to hide and tap **DONE** located at the bottom of the screen.

As shown below

4. To unhide the apps, just repeat the steps 1 to 3 above and unselect the apps you have selected before. Then tap **DONE** located at the top of the screen.

Managing Applications

You can force stop a misbehaving app. In addition, you can check the app data usage an also be able to uninstall the app.

1. Swipe down from the top of the screen and tap settings icon

2. Swipe up and tap **Apps**.
3. Tap an app to manage.
4. To force an app, tap **FORCE STOP.** Force stopping an app is useful when an app is misbehaving or refuse to close. To access a force stopped app again, just relaunch the app from the application screen.
5. To **Uninstall** the app, tap on Uninstall

As shown below

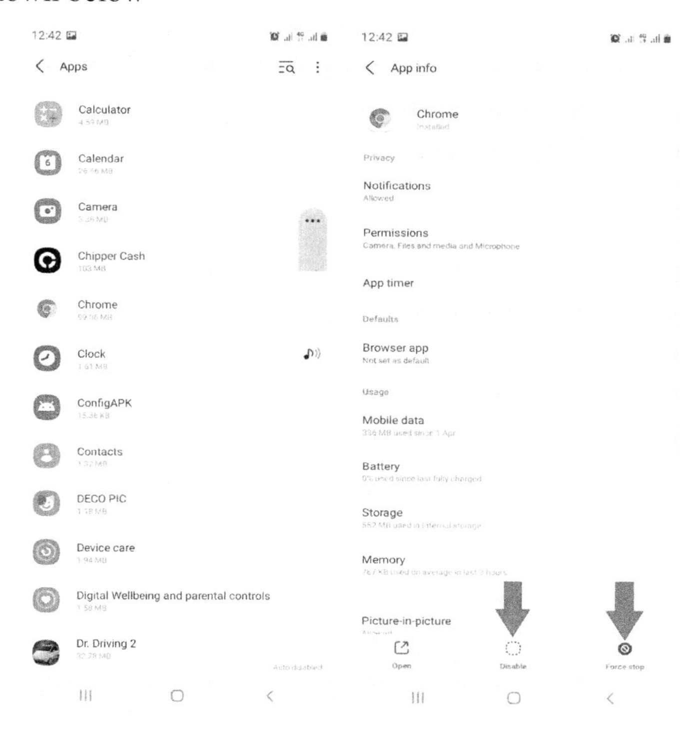

6. To manage the notification of an app, tap **Notifications** and choose an option.
7. To manage the permissions you give to an app, scroll down and tap **Permissions** and choose an option.

Managing Phone Notifications

It is good to know that Notifications consume battery and may be a source of disturbance sometime.

To manage notifications:

1. Swipe down from the top of the screen and tap settings icon

2. Then tap on the search icon.
3. Key in **Notifications** into the search bar. The result filters as you type. Tap **Notifications** from the results that appear.

As shown below

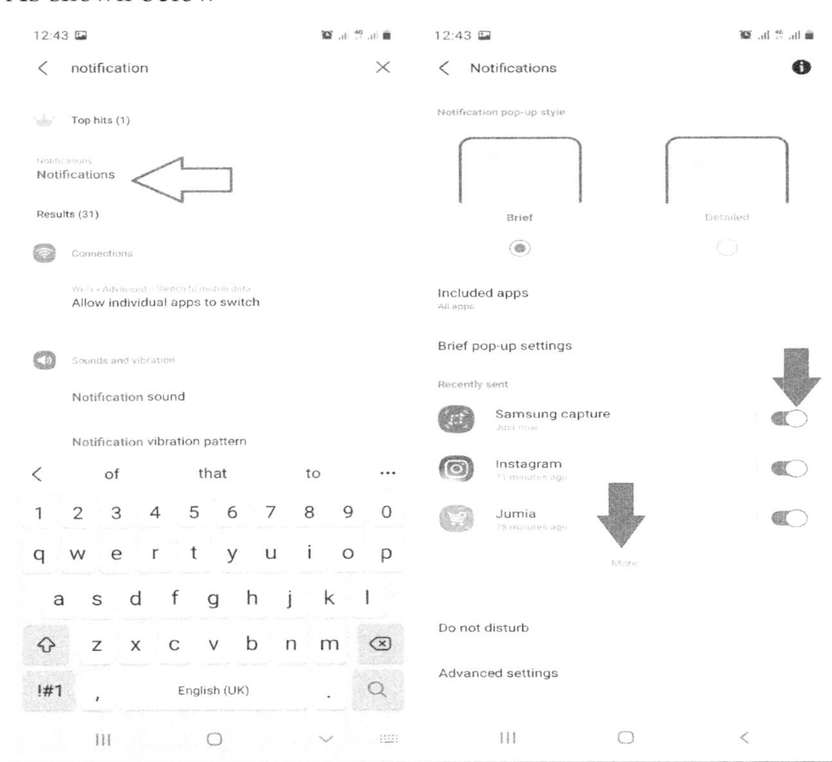

4. Tap on **See more** to check more apps, toggle to enable/disable any apps notifications.

Also under the **Notifications** you can enable/disable **App icon badges**.

- Tap on **App icon badges**

You can also customize the app icon badges by selecting any of the options.

As shown below

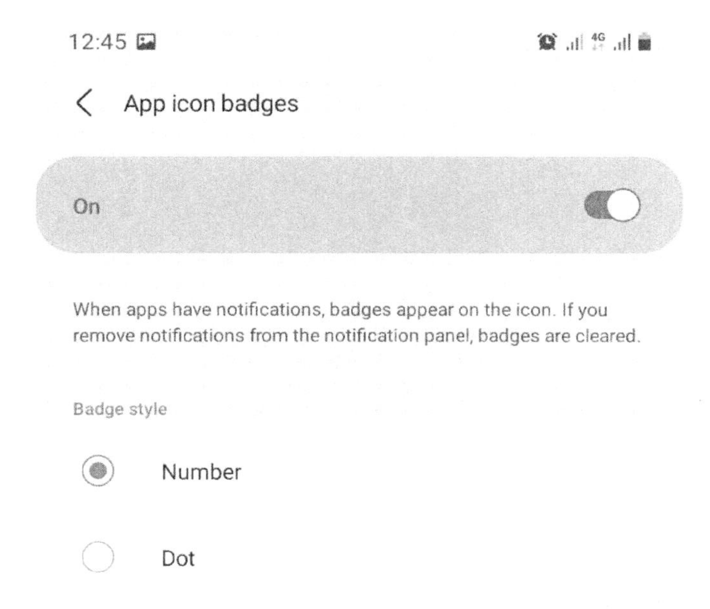

Note: If you are not interested in the app icon badges, toggle to disable the feature. Also note that whenever you remove notifications from the notification panel, badges are cleared.

Using the Multi Window Function

Multi window also referred to as split screen is one of the coolest features of Samsung Galaxy A12. Multi window allows you to put two apps side by side and even copy information from one app to the other.

Note*:* Not all apps support multi window feature.

Using Multi Window feature:

Before you start using split screen, It is advisable that you enable Recent button multi window feature. To do this:

1. Open the apps you want to use
2. Tap on recent button, and tap on the app icon

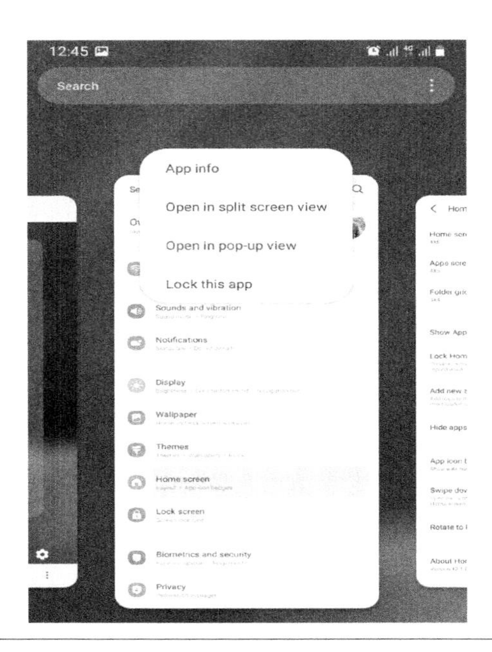

3. Tap on **Open in split screen view,** add another app.

As shown below

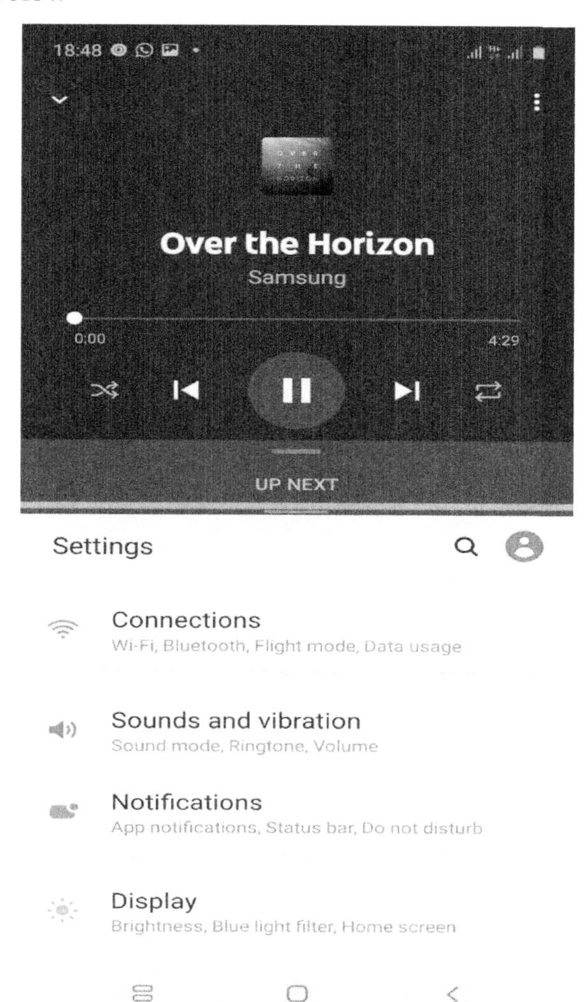

Note: You can adjust the screen size by dragging the upward or downward.

Quick Settings menu

The Quick settings panel present on notification panel provides quick access to device functions such as Wi-Fi, setting, data and more allowing you to quickly turn them on or off.

Tip: The app icon that is currently active in quick settings menu will appear bold and in color (blue), so if you want to know whether you have enabled a feature or not, just check it color/boldness.

As shown below

To view additional Quick settings:

1. Swipe down from the top of the screen to display the Notification panel.
2. Swipe down again to have the full view of the pannel.

Customize the Quick settings:

1. Swipe down from the top of the screen to using two fingers.
2. Tap the menu icon ⋮ located at the top right corner of the notification panel.

As shown below

3. Tap **Button order.**
4. Drag any of the icons to change their positions.

As shown below

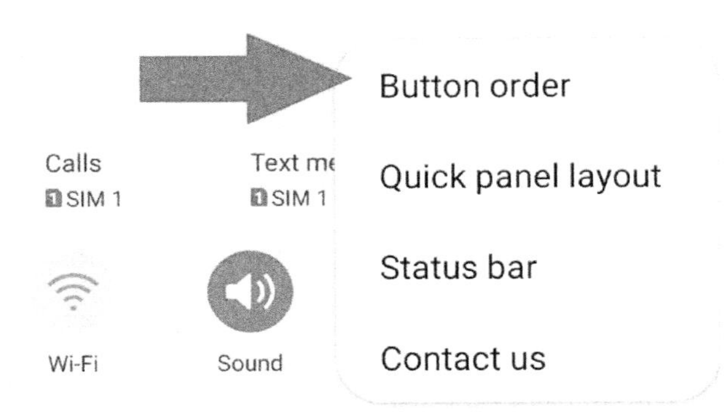

5. To add an app to the quick settings panel, drag the app from lower section of the screen to the top section of the screen.
6. To remove an icon from the quick settings panel, drag the app to lower section of the screen.
7. Tap on **Done**

Quick panel layout

If you tap on quick panel layout as displayed above, you will be able to make some tweaks on your panel. Such as: **Show brightness control above notifications and show multi sim info**. As shown below

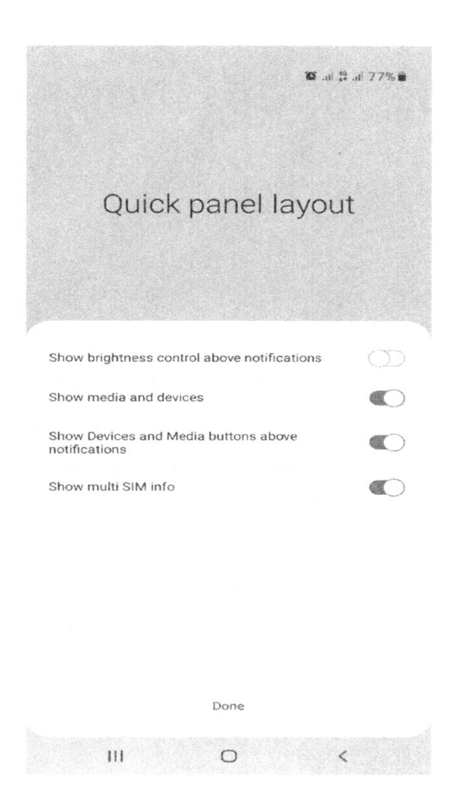

Status bar

If you tap on status bar as shown on page 42, you will be able to quickly access some of the advanced settings like **show battery percentage, app icon badges, notifications reminder** and other cool feature.

Customizing Your Phone

You can get more done with your phone by customizing it to match your preference.

To change your language

1. Swipe down from the top of the screen and select Settings icon , and scroll to **General management.**
2. Tap **Language and input.**
3. Tap **Language**
4. Tap on **Add language**

As shown below

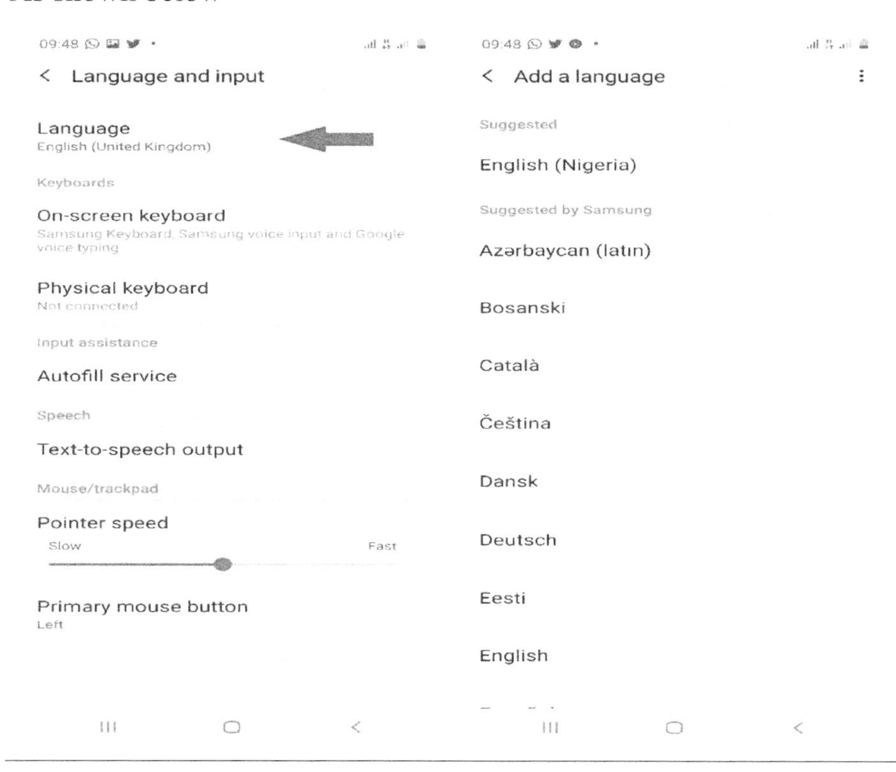

5. Tap and hold on any language you want to delete.
6. Tap on **Remove**

As shown below

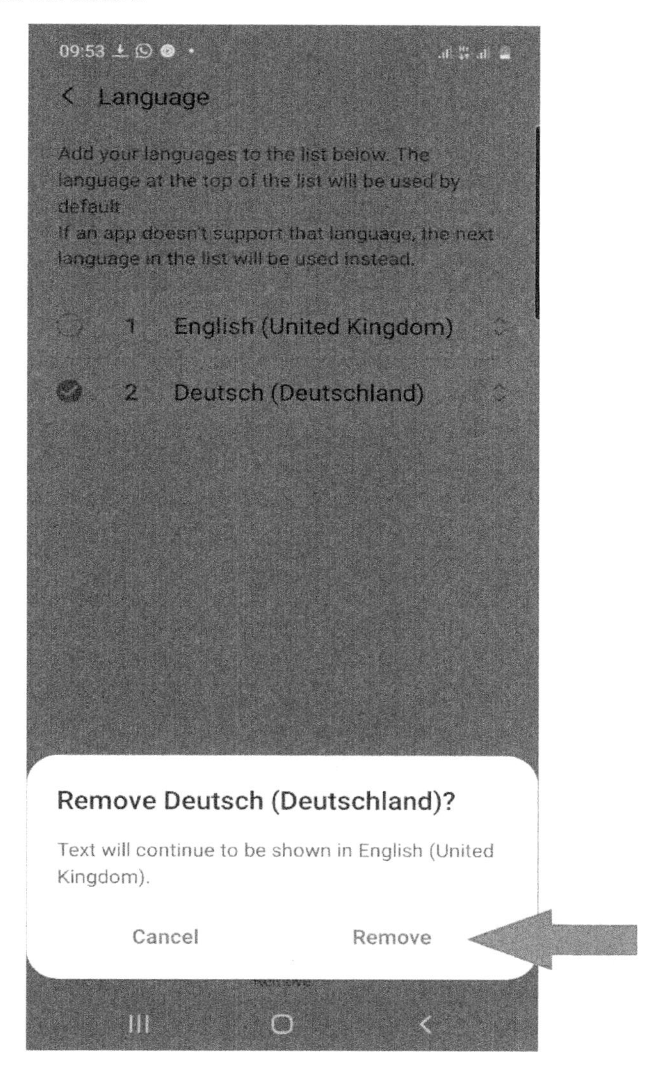

Selecting the Default Keyboard

1. Swipe down from the top of the screen and select Settings icon , scroll to **General management** and tap it.
2. Tap **Language and input.**
3. Tap **On-screen keyboard** and then choose a keyboard.

As shown below

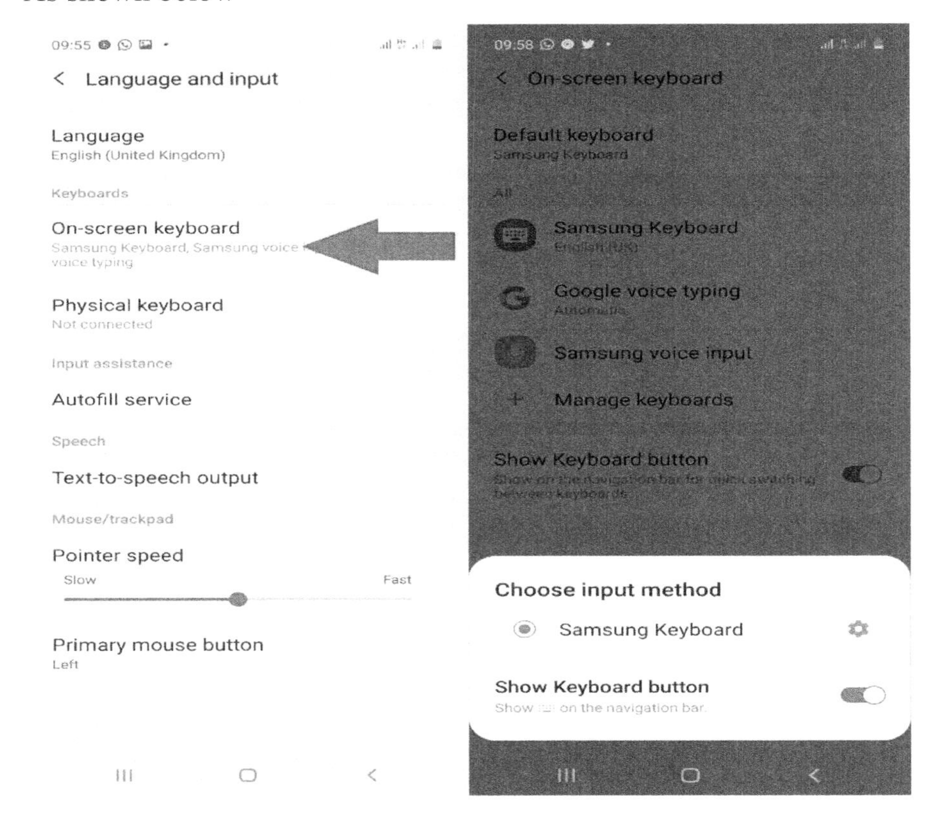

Note: If you have not downloaded extra keyboards from Google Play, you may just see only one option. Go to Google Play to download keyboard of your choice. Follow the same procedures to change your keyboard to the downloaded keyboard.

Set the current time and date

Your device is built to update its time automatically but you may need to manually set your time for one reason or the other.

To set time and date:

1. Swipe down from the top of the screen and select Settings icon , scroll and tap **General management** .
2. Tap **Date and time.**
3. Toggle to enable **Automatic date and time,** and **Use 24-hour format**.

Note: You can disable the Use 24-hour format if you don't like it. by default the feature is enabled.

Sounds and vibrations

1. Swipe down from the top of the screen and select Settings  icon .
2. Tap **Sounds and vibration.**

As shown below

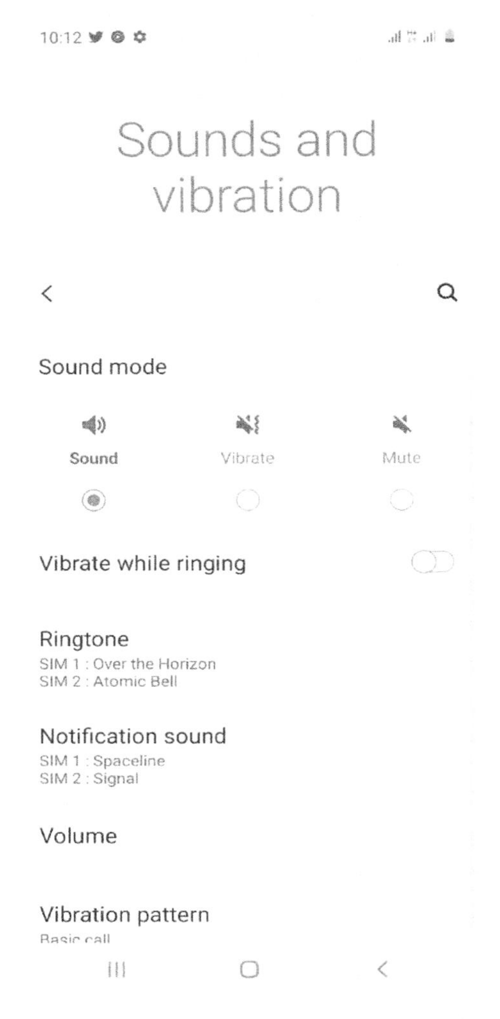

3. Tap on options to set your device sound and vibration

To adjust your volume

Press the **Volume key** up or down, this is the long key located at the left side of your Samsung Galaxy A12.

Adjusting the brightness of the display

1. Swipe down from the top of the screen using your two fingers. Then drag the slider under the quick action icons to adjust the brightness.

As shown below

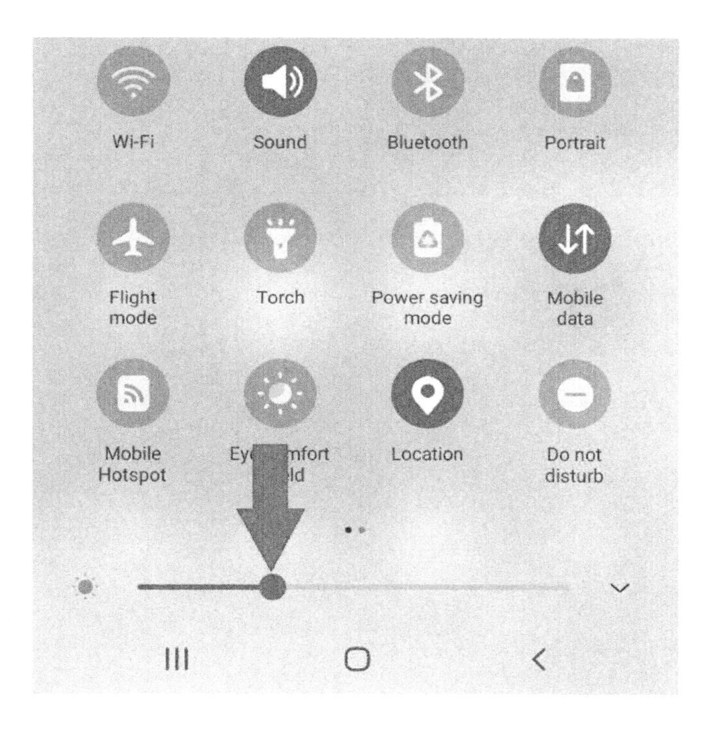

Note: The brightness level of the display will affect how quickly your device consumes battery power.

To set screen lock password or PIN

You can lock your phone by activating the screen lock feature

1. Swipe down from the top of the screen and tap Settings icon

    .

2. Tap **Lock screen and security**, Tap on **Screen lock type**
3. Tap a screen lock type you like.
4. If you choose **Password or PIN**, then enter the **password/PIN** you like and follow the onscreen instructions.

Note: You may choose **Pattern** or **Swipe.** If you don't want a lock screen, tap **None.** Also know you would be asked your password anytime you are trying pen your phone when locked.

Entering a text and More about keyboard

You can enter text by selecting characters on the virtual keypad or by speaking words into the microphone through the use of voice command.

Also you can do more with your keyboard by sending Gifs, Emoticons and other customized texts.

Keyboard Layout

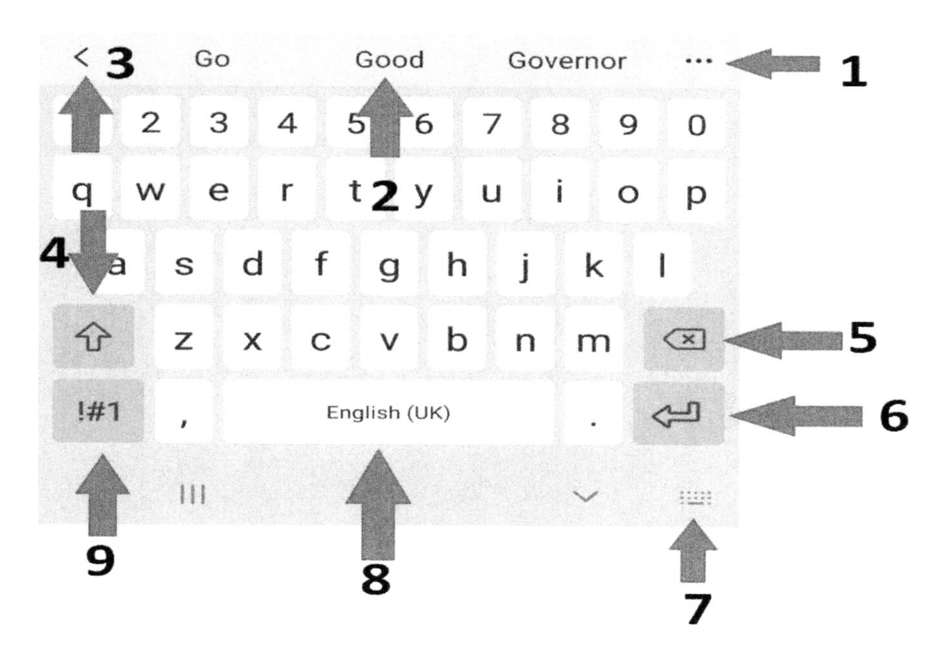

Number	Function
1.	Tap on this to see more predictive texts.
2.	**Predictive text bar:** As you type, your phone will give you text suggestions, the most likely suggestions will appear on the middle space in the predictive text bar.
3.	**Options bar**: Tap to see more options on your keyboard (such as Search, Translate, Clipboard and other options).
4.	Change case
5.	Clear your input/backspace
6.	Start a new line. At times, it serves similar function with Enter as we have it on PC.
7.	Switch keyboard shortcuts: Tap to switch from the available virtual keyboard.
8.	**Space bar**: to create space in between texts while typing.
9.	Switch between Number/Symbol mode and ABC mode

Note: By default, the following will display on top of your virtual keyboard.

1. **Emoticon**: Tap this to add emoticons to your texts. There different emoticons depending on your mood. This helps make chats more interesting.
2. **Stickers**: Tap this to create your personal emoji and to get more stickers. All this will definitely make your chats lively.
3. **Voice input**: Enter text by voice.
4. **Keyboard settings:** Tap this to access the keyboard settings.
5. **Options menu:** Tap on the menu bar to get more options

As shown below

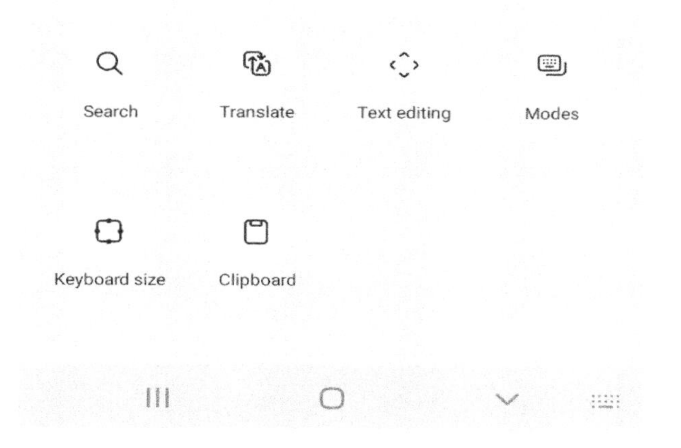

How to copy and paste a text

If you are using an app and you want to copy a link/word from another app to the particular app, this is how to copy and paste:

1. Tap and hold a word to display copy options. The icons below will show up after selecting a text or texts.

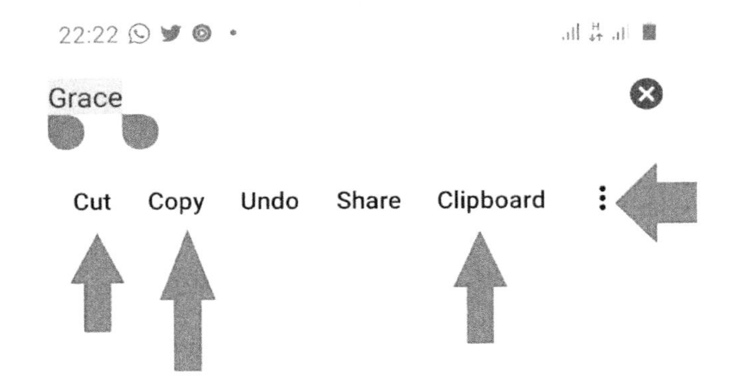

2. If the texts are much drag or to select more texts.
3. Select **Copy** to copy, or select **Cut** to cut the text. Also you can tap on the three dots icon to check out more options.
4. Go to the other app you intend to paste what you've copied, tap and hold on the text space.
5. Select **Paste** to insert the text you've copied.

Note: All recently copied items would be saved in the clipboard. Tap and hold on the text space then tap on **clipboard** to access more paste options.

How to use voice typing

1. Tap on voice input icon .
2. Speak your text. Your device types as you speak.
3. Tap on the voice input icon to **Pause**

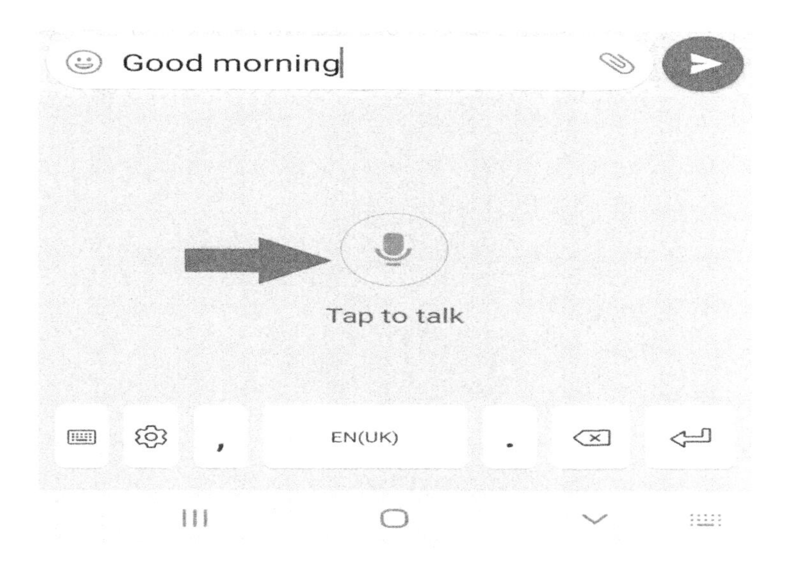

To customize the voice typing option:

- Swipe down from the top of the screen, and tap **Settings**

- Scroll, tap on **General management** and tap **Language and input**.
- Tap **On-screen keyboard** and tap **Google voice typing.** Then use the various options to customize it.

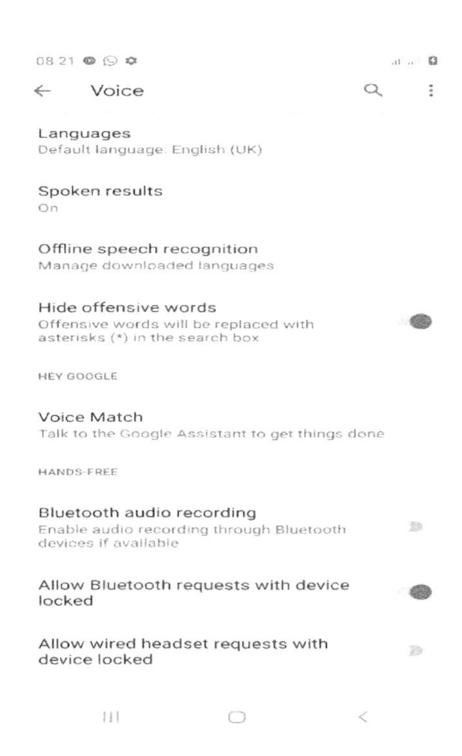

Using the Fingerprint Features

One of the interesting features on Samsung Galaxy A12 is Fingerprint. It allows you to unlock applications without going through the stress of entering your passwords.

Note: A12 fingerprint is located right on the power key button.

How to register your Fingerprint

You will need to first register your fingerprint before you can use it. You have the chance of registering more than one fingerprint.
Note: You will be asked to set password that will serve as a backup for the fingerprint in case your device fingerprint sensor misbehaves.
To do this:

1. Swipe down from the top of the screen and select Settings icon ⚙ .
2. Tap **Lock screen,** then tap on **Screen lock type**
3. Tap on **Fingerprints** under the **Biometrics**.
4. Tap **Continue**
5. Place one of your fingers on the fingerprint reader located on the power key of the phone, place and adjust your fingers on the power key then lift it when the fingerprint is detected and read. You may need to repeat this for a number of times.

Easy way to rename Fingerprints

1. Swipe down from the top of the screen and select Settings icon  .
2. Input **Add Fingerprint** in the search apace. Also you can go to **Biometrics and security**, then tap on **Fingerprints**. You will have the same result.
3. Tap on the result, and input your password to unlock.
4. Tap the fingerprint you want to rename. For example, tap **Fingerprint 1.**

As shown below

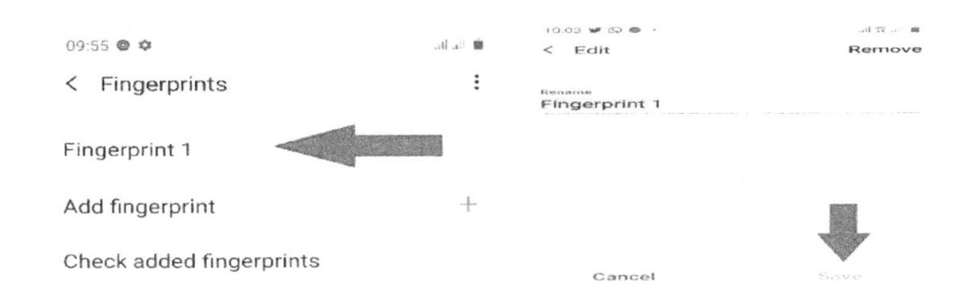

5. Enter a new name, and then tap **Save**.

How to delete Fingerprints

1. Swipe down from the top of the screen and select Settings icon  .
2. Tap **Biometrics and security**
3. Tap **Fingerprint**.
4. Unlock the screen using your password.
5. Touch and hold the fingerprint you want to delete, and then tap **Remove**.

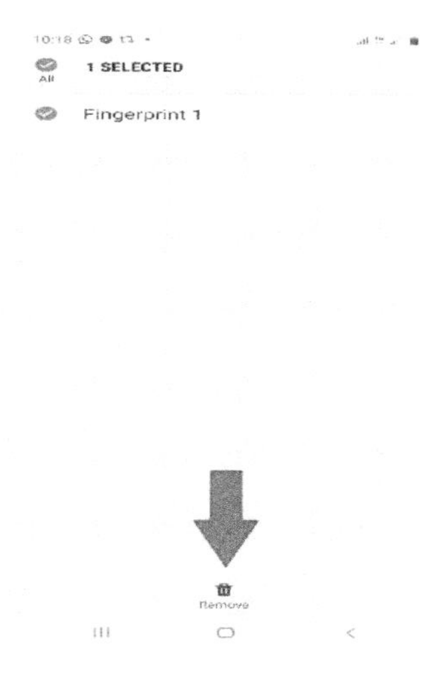

Note: Apart from using your fingerprint to unlock your device, you can also use it to access online features, and sign into accounts. Such as Samsung account and some other online accounts.

Using the Edge Panels on Samsung Galaxy A12

The edge panels makes the use of your Samsung Galaxy A12 more interesting. Edge panels can be used to access apps, tasks, contacts and more. Edge panels is same thing as Edge screen.

To access edge panel, swipe the edge panel handle located at the edge of the screen. To access more edge items, swipe again from the edge of the screen.

As shown below

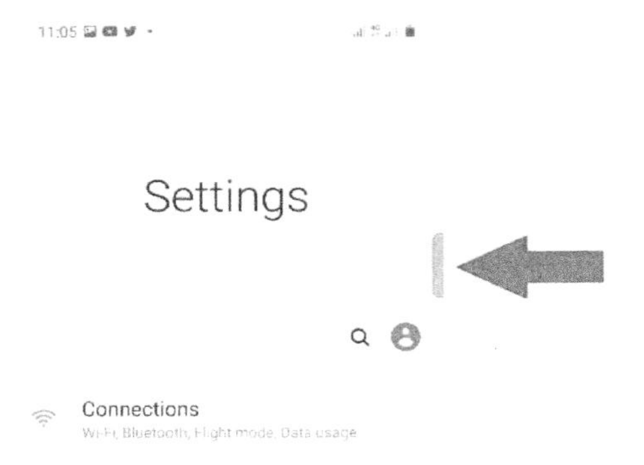

To enable this feature:

- Go to **Settings**
- Tap on **Display**, scroll and tap on **edge panels**.

As shown below

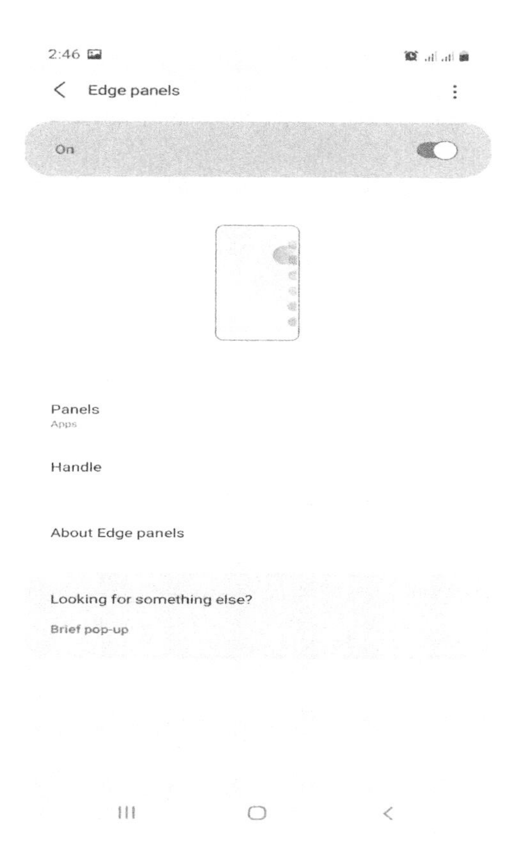

Note: After you have enabled this feature, you can access the **edge panels** settings when you tap the setting icon located at the bottom right corner of the screen.

As shown below

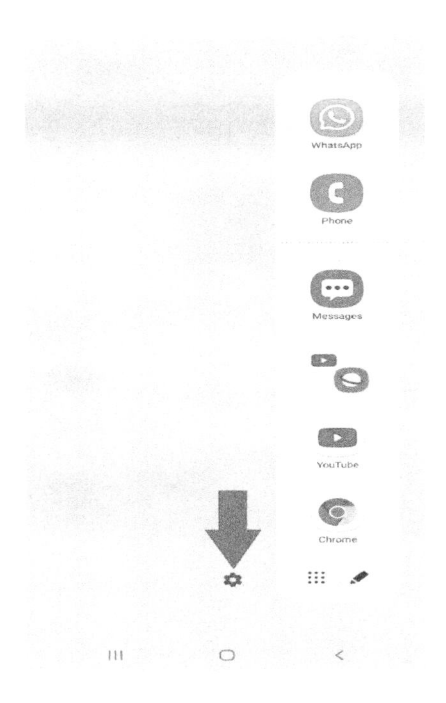

Using the Edge Panels

The Edge Panel allows you to use the edge screen in a special way.

To manage the edge panel:

1. Swipe down from the top of the screen and select Settings icon , and then tap the **Display** and tap on **Edge panels**

OR

2. Swipe from the edge of your device to access **Edge panels**

Note: You can tap on Galaxy Store to download some exciting panels.

4. To uninstall an edge panel that you don't want again, tap the menu icon ⋮ located at the top right corner of the edge panel's screen and select **Uninstall**, then select the minus icon (--) located at the top of the panel you want to uninstall.
Note: The uninstall button may not be available if you have not installed any handle panel from edge panel store.

As shown below

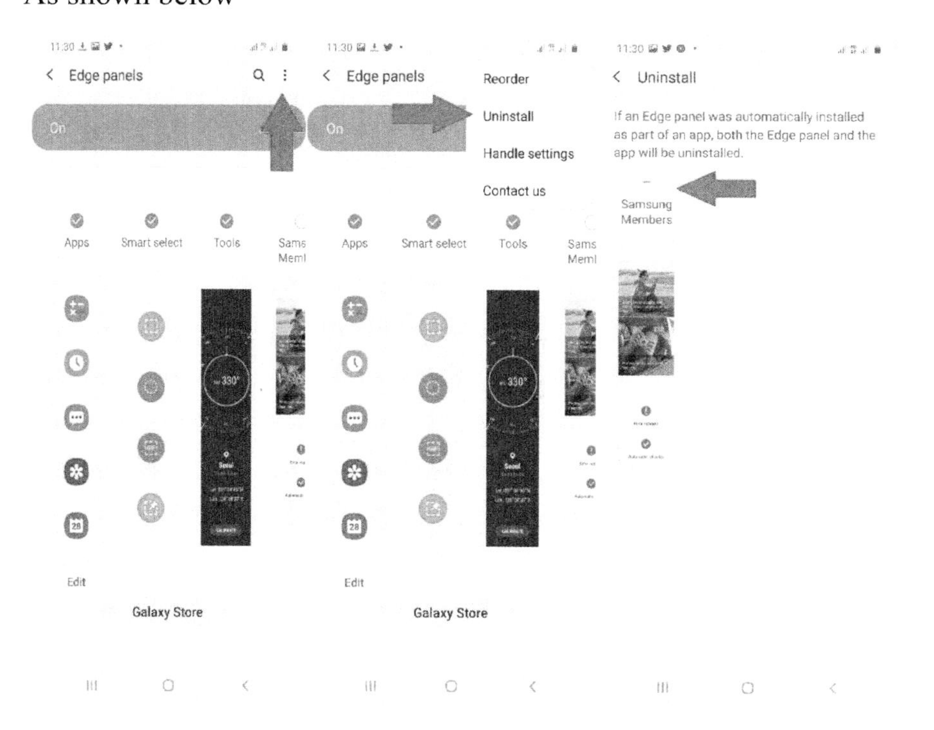

How to Edit the Edge Panels

There are some edge panels that allow you to edit them.

To edit panels that allow editing,

- Tap on **Edit.**

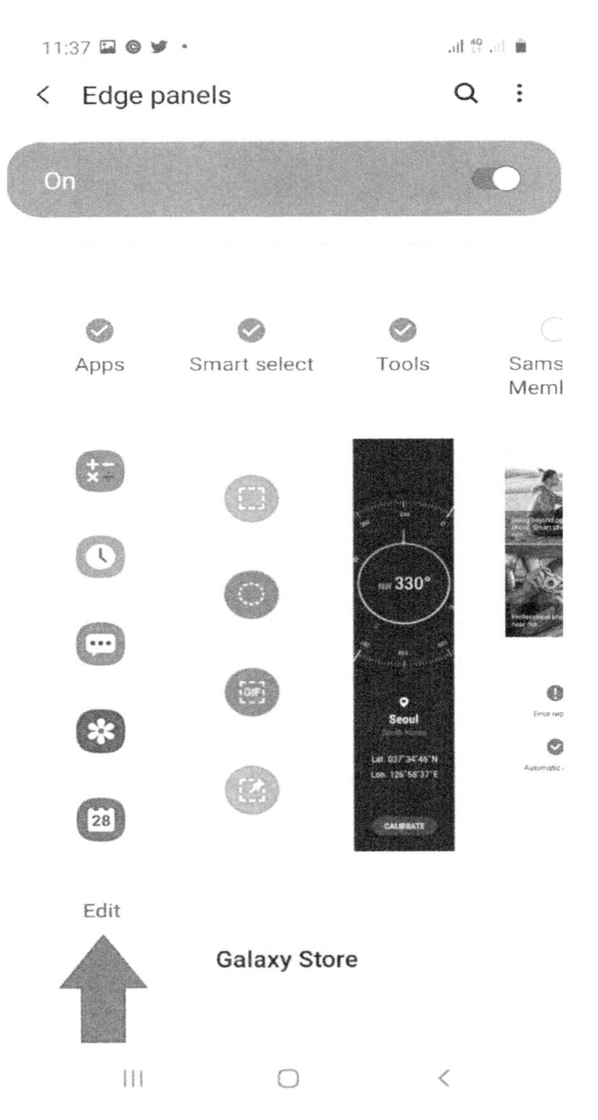

- Then tap and hold an icon and move it to the edge screen. To remove an item, tap the minus icon. To Reorder the icons, tap and drag the icon(s) to a new location.

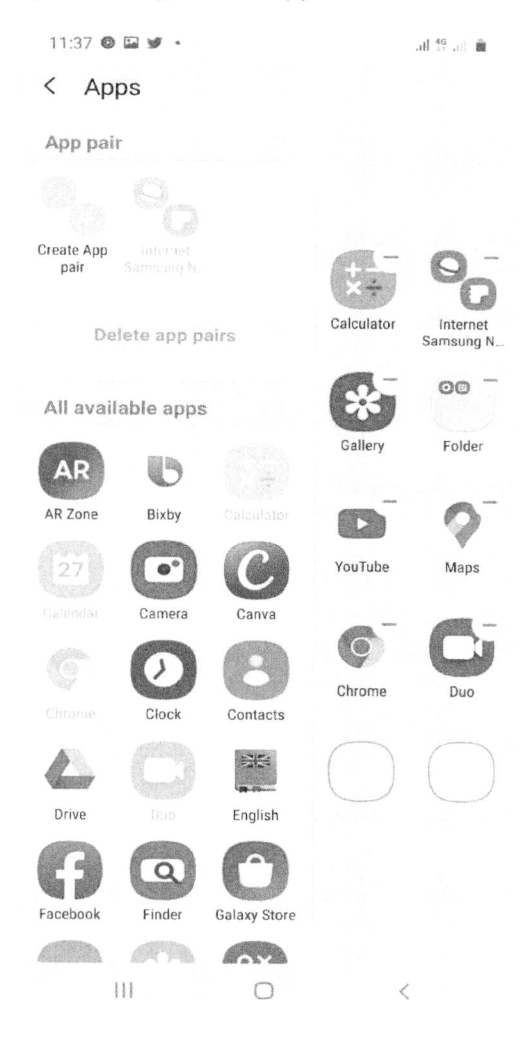

Note: that there is a maximum number of items you can add to the edge screen.

- To select the edge panels that you want to see on the edge screen, just tap to mark the circle on top of the panel.

As shown below

- To disable edge panel completely, go to **settings**, tap on **display** and toggle **edge panels** to disable.

Note: Anytime you disable Edge Panel, the edge handle will disappear.

Using the Internet App

Opening the Internet browser

To access internet browser

- Tap on the **Internet** .

Note: There are different internet browser. Such as Google Chrome,Opera Mini and others.

- Tap **Internet** . On Home Screen If you are using the Internet app for the first time, then follow the instructions.

Get to know the Internet Browser Layout

The following are the internet browser icons:

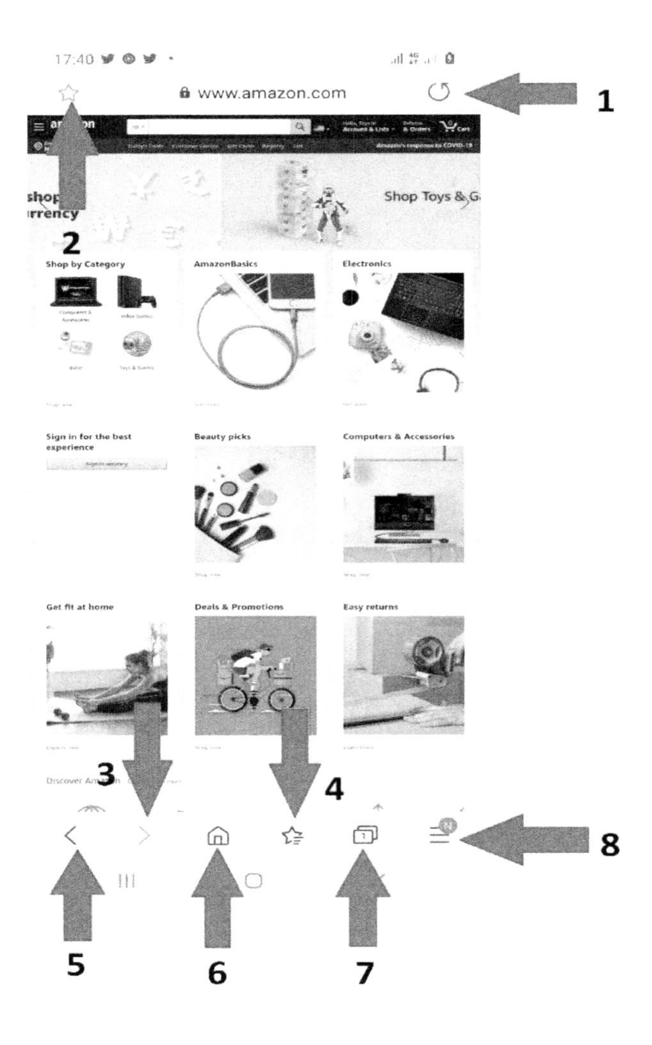

Number	Function
1.	**Refresh:** Tap this icon reloads a webpage.
2.	**Bookmark:** Tap this icon to bookmark. This icon will be marked in yellow color.
3.	**Forward**: Tap this icon to return to the page you just left.
4.	**Bookmark log:** Tap this icon to check the history, and the pages you have bookmarked.
5.	**Back:** Tap this icon to revisit the most recent page. To quickly access browsing history, tap and hold the back button < then tap on **History**.
6.	**Home:** Tap this icon to go to the browser's home page. The default home page is Google. Tap and hold the Home icon and then select **Other.** Then enter the website of your choice (e.g. Amazon.com) and tap **OK** to add web sites to your browser home.
7.	**Tabs:** Tap this icon to navigate between different webpages you have opened.
8.	**Menu icon:** Tap this icon to access additional options such as **Downloads**, **History** and others. As shown below

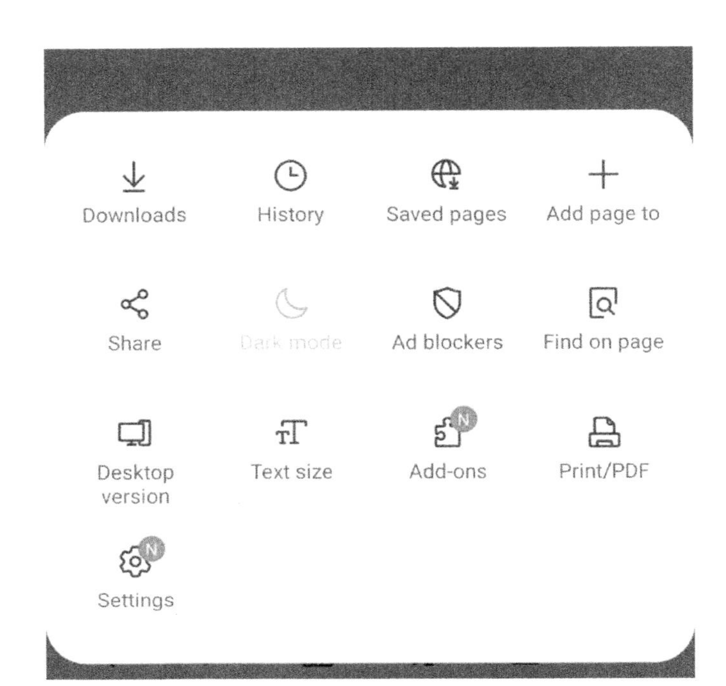
Downloads
History
Saved pages
Add page to
Share
Dark mode
Ad blockers
Find on page
Desktop version
Text size
Add-ons
Print/PDF
Settings

Phone

Learn how to use your phone app. You can use it to save contacts, make calls and some other call related features.

To make a call or silence a call

- While on the home screen, tap **Phone** and enter a phone number. If the keypad does not appear on the screen, tap the **keypad** to show the keypad [You may have a single on your device] . To call a number on your contact, tap the **Contact** button on the Phone app screen. And to call a recent dialed number, tap on **Recent**.

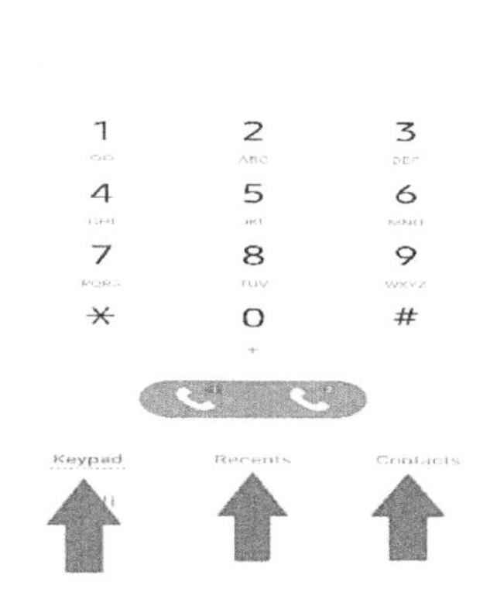

- To make a phone call, tap a contact and then tap the phone icon  .

To answer/reject an incoming call:

1. To receive an incoming call, tap the **Green phone icon** .
2. To decline an incoming call, tap the **Red phone icon**  .
3. To reply with a text, swipe up from the bottom of the screen and then follow the prompt instructions.

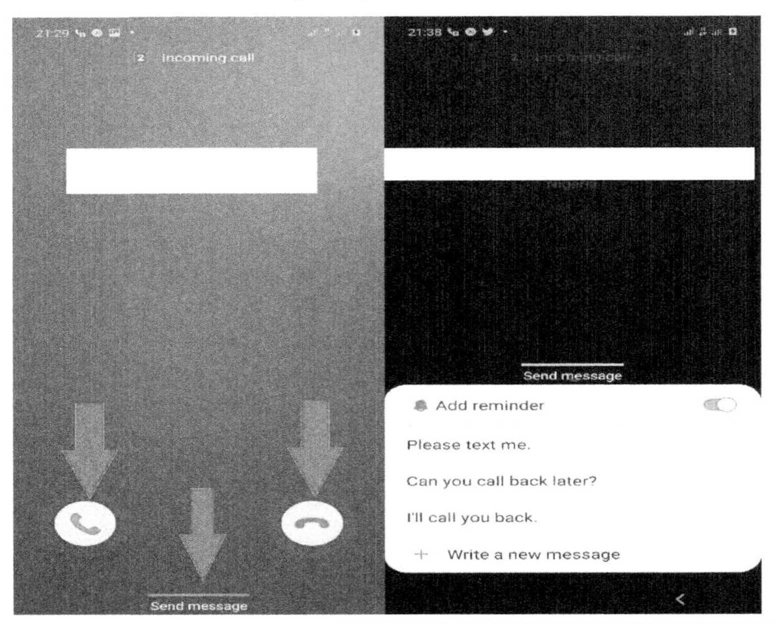

Phone Setting

- While on the home screen, tap **Phone** 
- Tap on three dots icon located at the top right corner.

As shown below

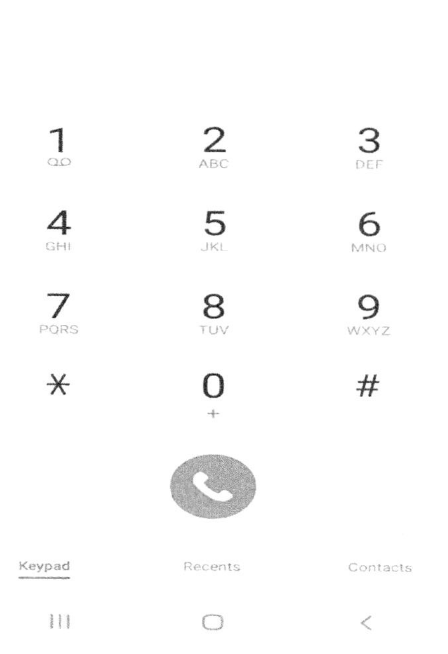

- Tap on **Settings**

As shown below

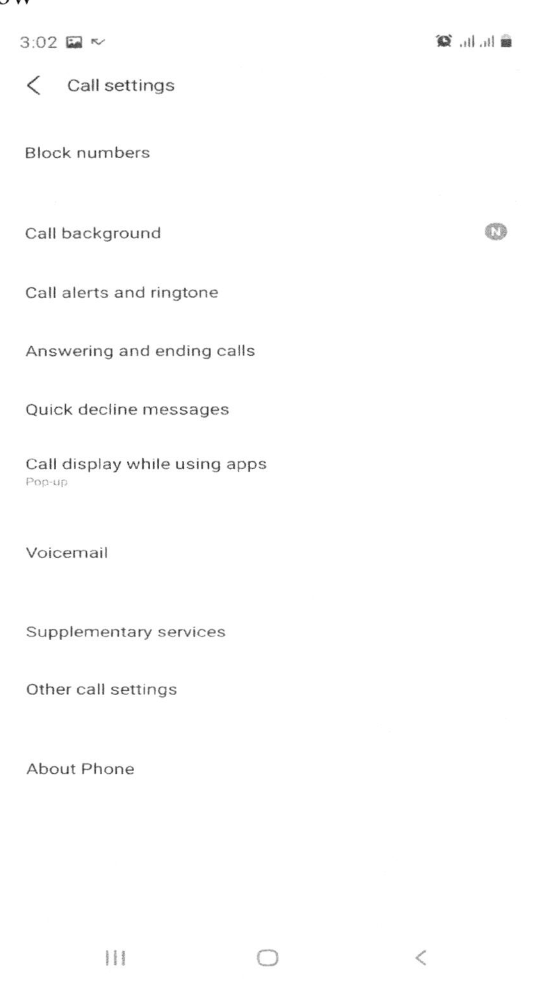

Block numbers

This allows you to block calls and messages from specific phone numbers.

- While on the home screen, tap **Phone** 
- Tap on three dots icon located at the top right corner then tap on **Settings**.
- Tap on **Block numbers**

Here you can block a number straight from the recent calls you made, straight from the contact or by dialling the number.

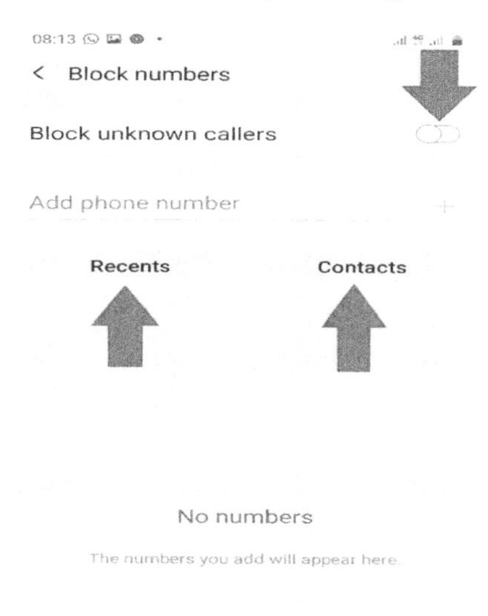

- Toggle on **Block unknown callers**, to block unsolicited calls from unknown people.

Call alerts and ringtones

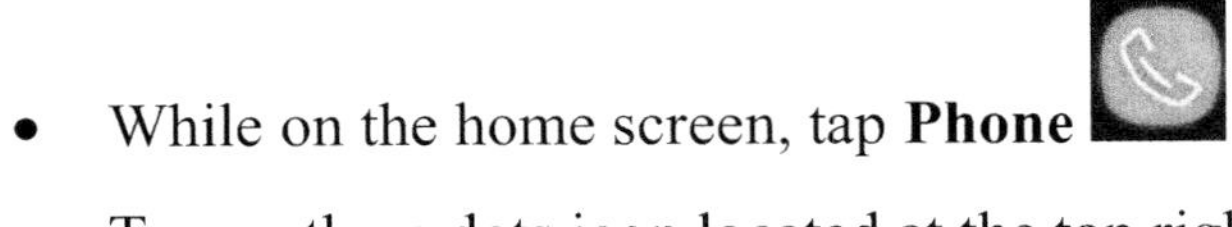

- While on the home screen, tap **Phone** 
- Tap on three dots icon located at the top right corner then tap on **Settings**.
- Tap on **Call alerts and ringtones**.

As shown below

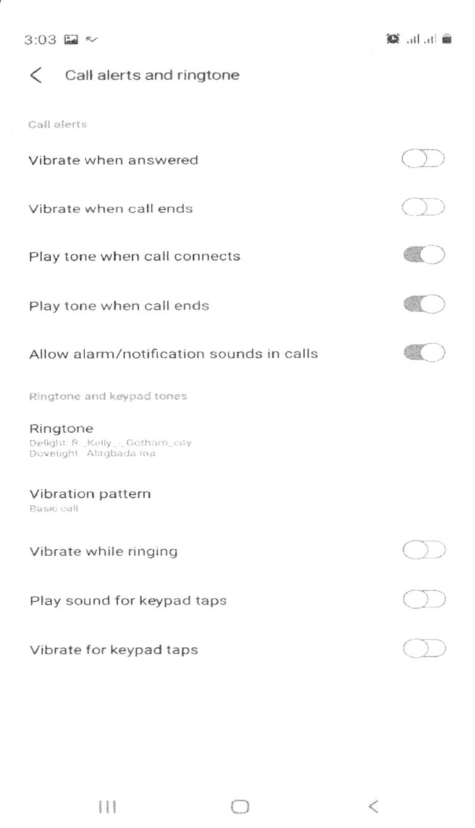

Here you will be able to make some tweaks to your call alerts and ringtones settings.

Answering and ending calls

This allows you to to use the volume key to answer and end calls.

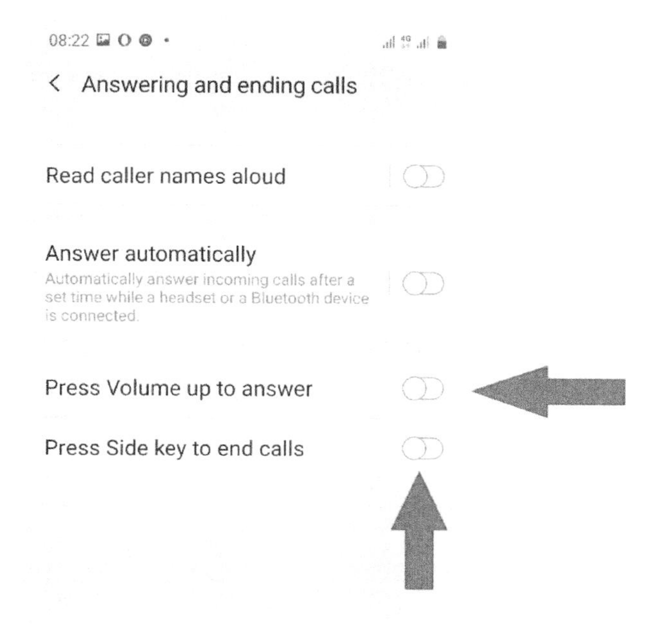

- While on the home screen, tap **Phone**
- Tap on three dots icon located at the top right corner then tap on **Settings**.
- Tap on **Answering and ending calls**
- Toggle to enable Press Volume up to answer and Press Side to end calls.

Set Quick Decline Messages

This allows you to add and remove quick decline meassages.

As shown below

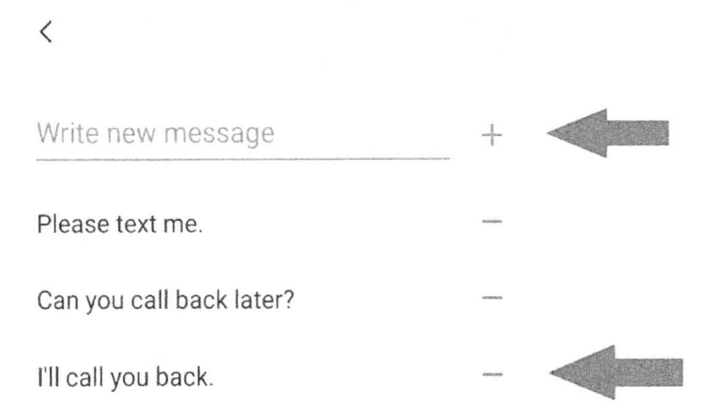

- Tap on the Plus (+) icon to add after typing.
- Tap on Minus (–) to removed the already added message(s).

Note: The decline messages would appear anytime you reject/decline calls, then you select from the messages.

Call display while using apps

- While on the home screen, tap **Phone** 
- Tap on three dots icon located at the top right corner then tap on **Settings**.
- Tap on **Call display while using apps**.

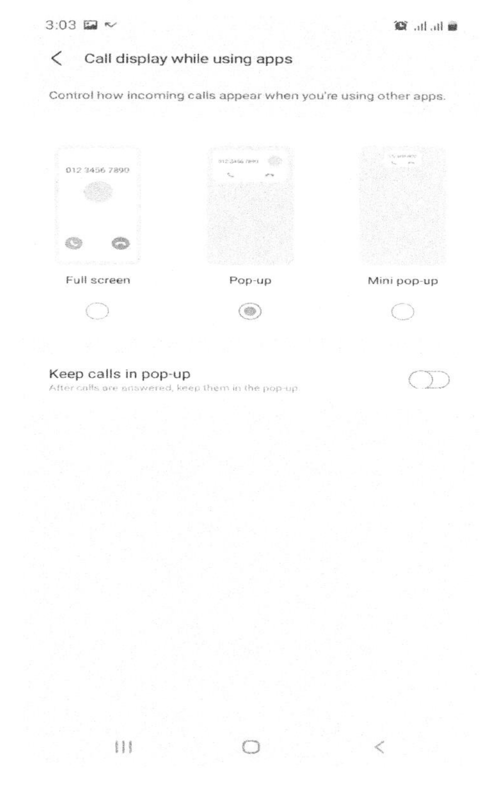

- Tap to select any of the options.

Note: This feature controls how incoming calls appear when you are using other apps.

Using the speed dial options

The speed dial allows you to quickly access a number in your contact.

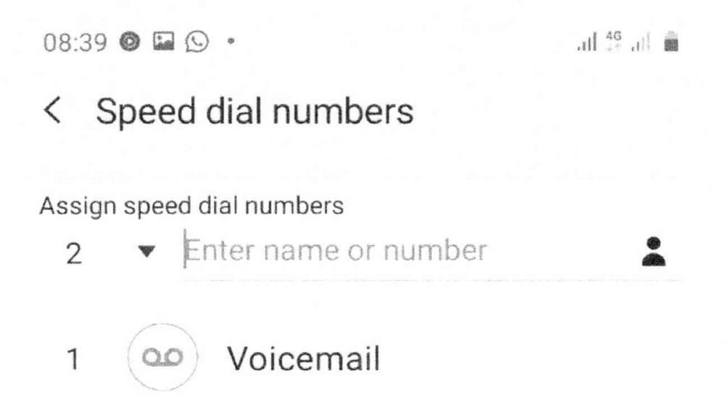

To get this:

1. Tap the phone icon
2. Tap **menu icon** ⋮ located at the top of the screen.
3. Tap **Speed dial**.
4. To add a contact to speed dial, tap contact icon to add a contact.
5. To change the speed dial number, tap the arrow icon and pick a number. Number one is reserved for voicemail.

6. To call a contact you have added, tap the phone app icon and from the dial pad, press and hold the number assigned to the desired contact. To place a call for a contact assigned to a two-digit speed dial, dial the first number, then press and hold the 2nd number.
7. To delete a speed dial number, tap the **Minus (-)** icon next to the assigned contact. Removing a contact from a speed dial list will not delete it from your phone.

How to use the Messaging app

This app allows you to send text, photo, and video messages to other SMS and MMS devices.

To start or manage a conversation:

1. Tap the **Messages** app icon from home screen.
2. Tap the new message icon located at the bottom right corner of the screen.
3. Type in the first letters of the recipient's name. Your device filters as you type. Then tap the required contact.

Note: You can add up to 20 contact (if not more). If you don't have the number in your contact depending on your service provider,
To remove a contact from the send list, just tap the minus icon (-) next to the contact.
As shown below

4. Tap the text space to write your message.
5. To attach a file such as audio, pictures, tap any of the icons + and follow the prompts.
6. When you are done, tap icon
7. Tap and hold message text to forward and share a message.
8. Tap and hold the message text to **Delete** the message.

Message Settings

To go to Message settings:

1. Tap the **Messages** app icon from home screen.
2. Tap on the three dots icon.

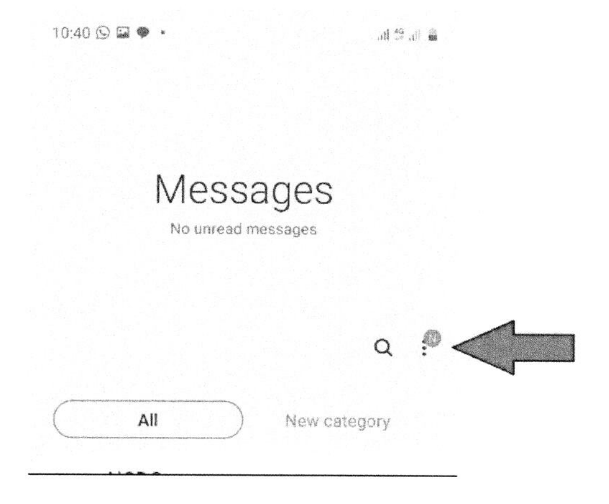

3. Tap on **Settings**.

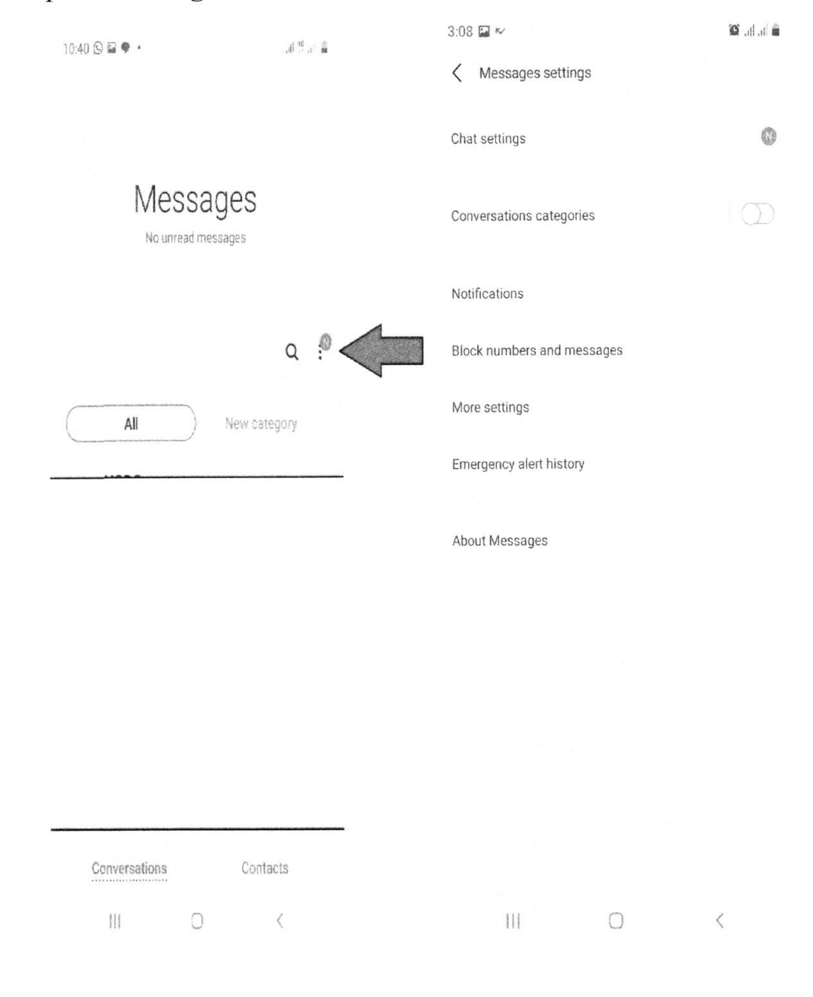

You can make some tweaks on your message settings.

Contacts

This app allows you to create, manage and save a list of your personal or business contacts. You can save names, mobile phone numbers, home phone numbers, email address, and more.

To create a contact:

1. While on the home screen, swipe up and tap on  **Contact** app.
2. Tap on **Add contact** icon located at the lower right corner of the screen.
3. Tap to select where you wish to save the contact.

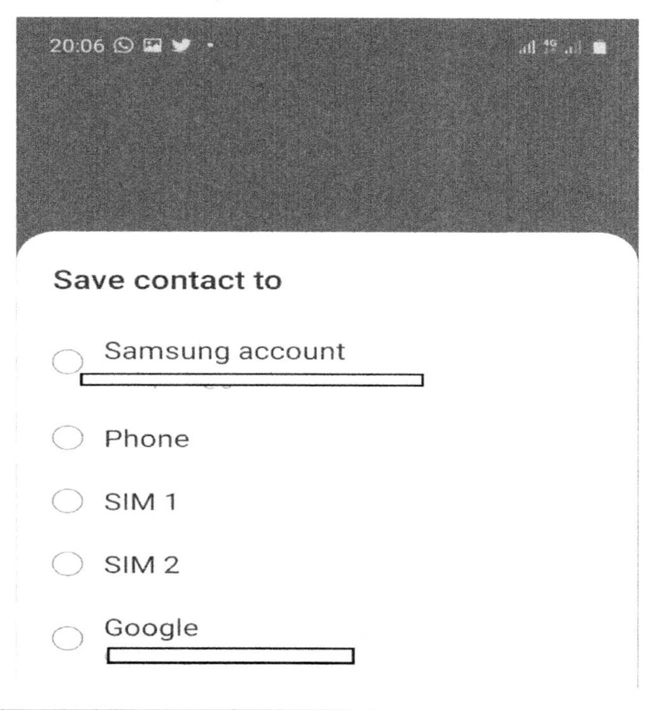

4. Fill in the details by tapping on each option on the screen.
5. To assign a contact to a group, tap **Group.**
6. To access more options, tap on **View more**.

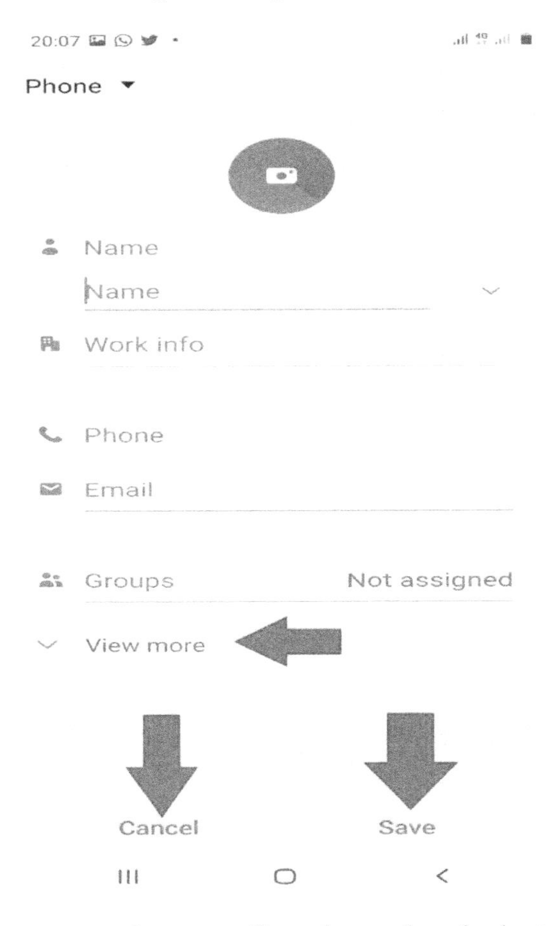

7. When you are done, tap **Save** located at the bottom right of the screen. OR you can tap on **Cancel** if you are no longer saving the contact.

Note: To easily search for a contact, open the contact app and tap **search** bar located at the top of the screen, and type the contact name.

To manage a contact:

1. Tap on **Contact** app.
2. Tap on a contact from the list.
3. Tap on **Edit** to edit the contact.

4. Tap on **Share** to share the contact with friends.
5. Tap on the three dots icon located at the top right corner to check more options. You can **Delete**, **Add/removed linked contacts**, and **Block contact** here.

Import or export contacts

If you have some contacts stored on your SIM card, you can move them to your phone or emial vis-à-vis.

How to do this:

1. Tap on **Contact** app.
2. Tap menu icon.

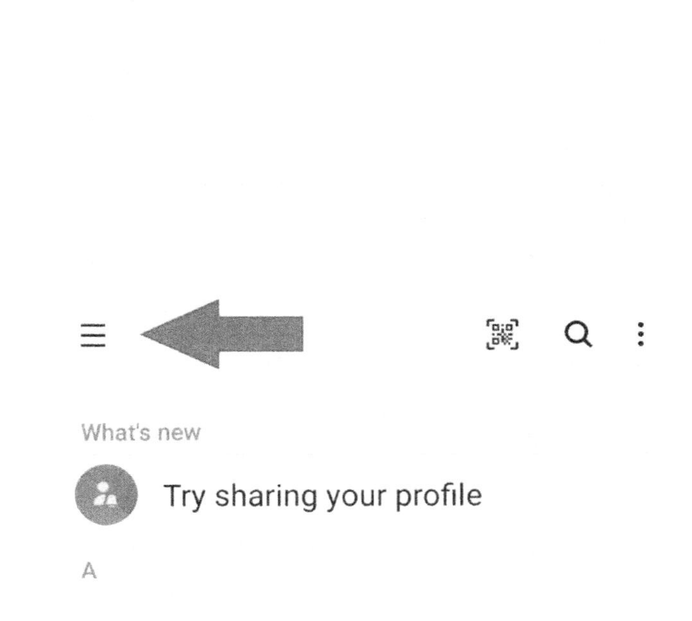

3. Tap **Manage contacts**.
4. Tap **Import/Export contacts**.
5. Tap **Import**.

6. Select where you are copying the contacts from.
7. If you want to import all the contacts, tap **ALL** located at the top of the screen.
8. Tap **DONE.**
9. Tap **Phone** or where you are sending the contacts to.
10. Tap **Import**

Do Not Disturb

This feature allows you to mute all calls, alerts, and media, except for your selected exceptions. Also it gives the opportunity to prevent unnecessary disturbances from your phone.

To quickly access this feature:

- Swipe down from the top of the screen using two fingers and then tap and hold **Do Not disturb.**

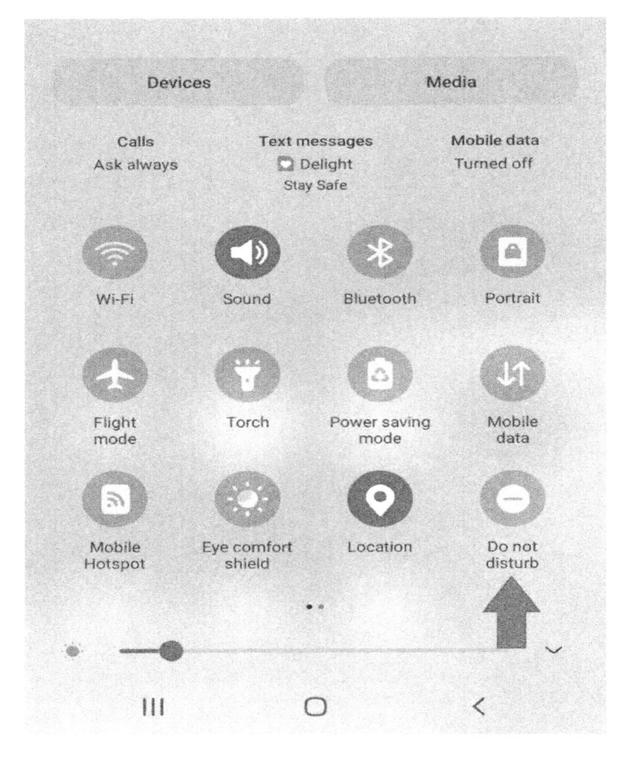

- Swipe to the left if you can't find **Do not disturb** on the first screen.

- Toggle to enable the feature. If you choose to schedule this feature, tap on **Add schedule**.

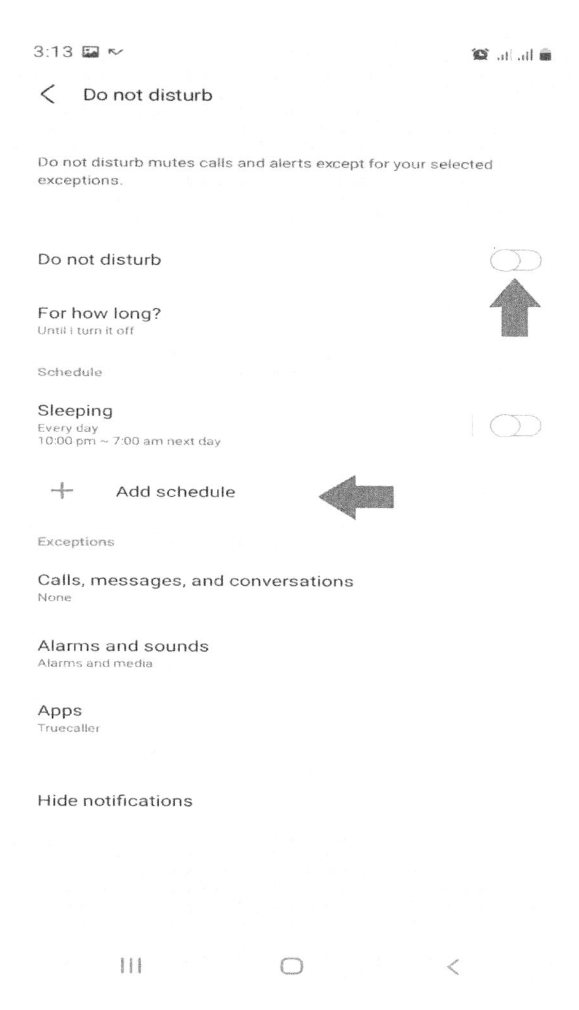

You can follow other options to customize your **Do not disturb.**

Connect your device to PC

Your phone can be connected to a PC with USB cable. This will enable you to transfer files such as audio files, document files and image files to your phone from your PC.

Note: You may need to buy a USB cable that works for your PC because the USB cable that comes with your Samsung Galaxy A12 might not work directly with your PC.

Warning: Please not disconnect the USB cable from a computer while the device is transferring or accessing data. This may result in data loss or damage to your phone.

Transferring file with USB

1. Connect your device to a PC with an appropriate USB cable.
2. When prompted to allow access to phone data, tap **Allow.** If you don't choose Allow, your PC may not see your phone.

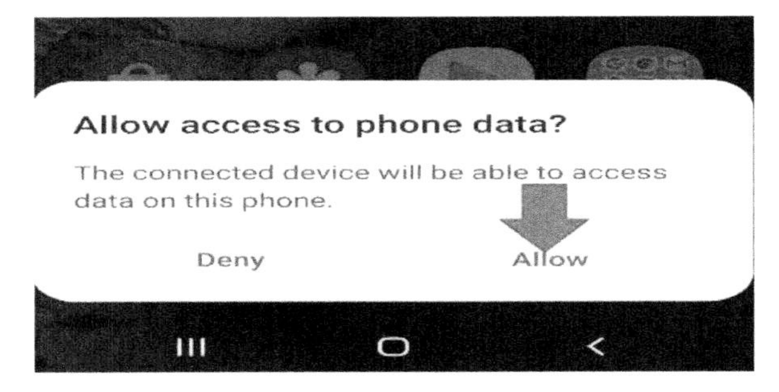

3. Also you can swipe down from the top of the screen and tap the USB option **(transferring files…)**.

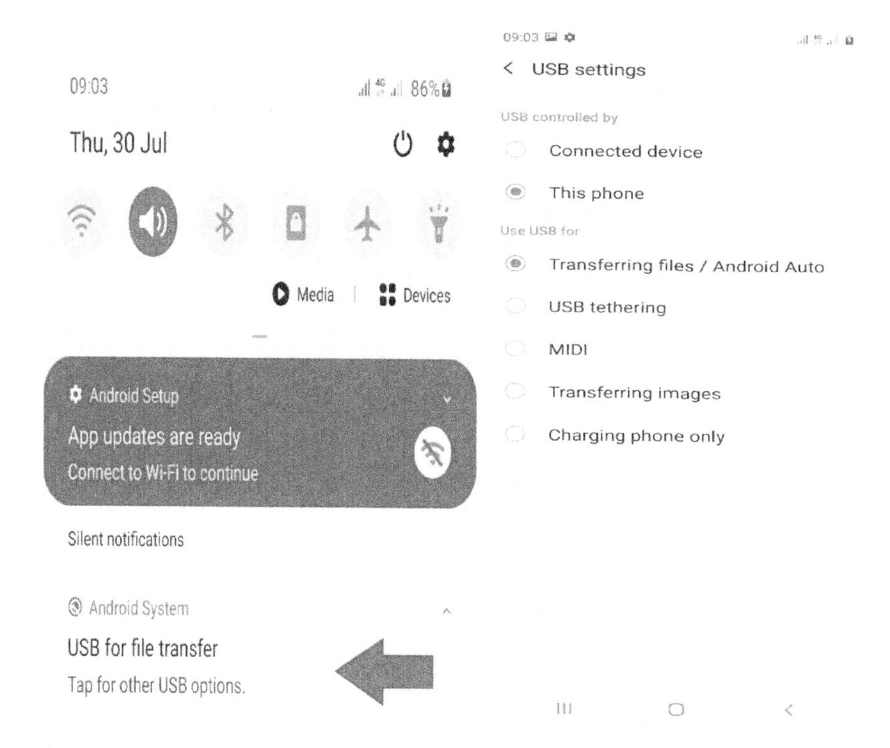

4. Tap **Transferring Files**. As shown above.

Your Samsung Galaxy A12 should appear in the same location as external USB drives usually appear. For Windows users, this is typically under "This PC/Computer" menu.

5. Open your device's drive so as to see the different folders present.

Note: You may not be able to access the folders if your phone is locked.

Connectivity

Wi-Fi

Using your phone, you can connect to the internet or other network devices anywhere an access point or wireless hotspot is available.

To activate the Wi-Fi feature:

1. Swipe down from the top of the screen.
2. Tap and hold the **Wi-Fi** icon.
3. Tap the switch next to Wi-Fi to turn it on.
4. Your device then automatically scans for available networks and displays them.
5. Select a network and enter a password for the network (if necessary). If connected the icon will be displayed on your screen.
6. To turn Wi-Fi off, swipe down from the top of the screen,

   Tap **Wi-Fi**  .

Mobile Hotspot

If your network provider support it, you can use this feature to share your mobile network with friends.

1. Swipe down from the top of the screen.
2. Tap on **Mobile Hotspot**.
3. Tap and hold **Mobile Hotspot** to access its' settings.

Here you will be able to set the password and do some other settings to secure your network.

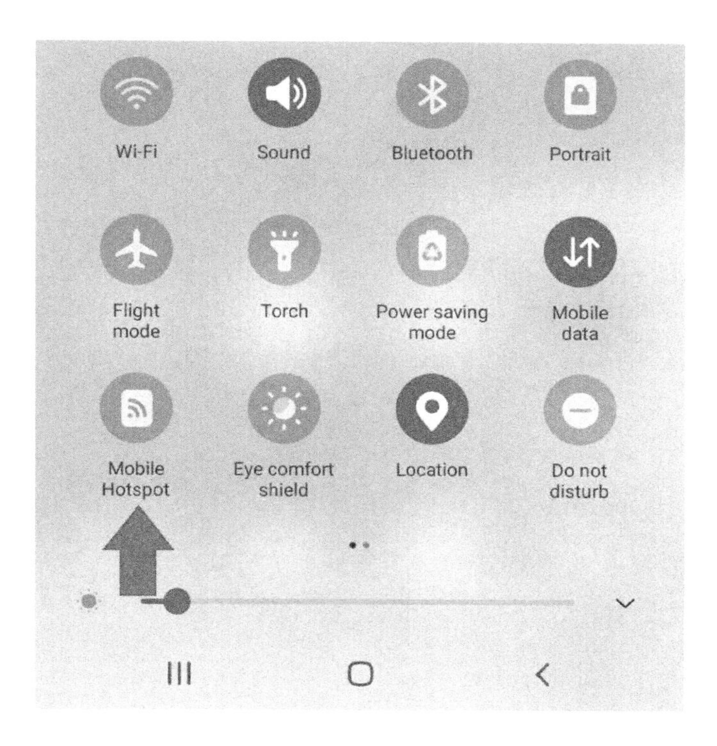

Using Bluetooth

1. Swipe down from the top of the screen.
2. Tap on **Bluetooth** .
3. Tap and hold **Bluetooth** to access its' settings.

Here you will be able to pair/connect with other device or PC and share files, audio with friends.

Location

Enabling location service allows Google Map, Google apps and some apps to serve you content and make your location visible.

1. Swipe down from the top of the screen.
2. Tap on **Location**.
3. Tap and hold **Location** to access its' settings.

Here you will be able to do some other settings.

Camera and Video

Using the Camera

Both Samsung Galaxy A12 come with rear-facing camera, front-facing camera and LED flash. With these cameras, you can capture a photo and record a video.

Camera layout

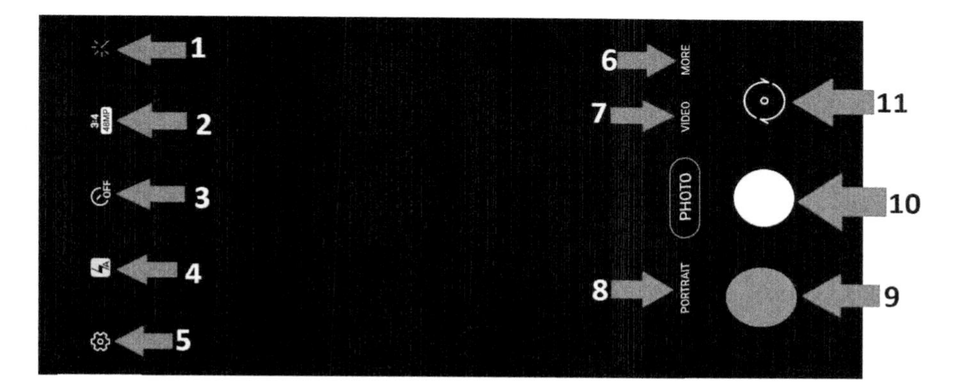

Number	Function
1.	Animated filter and Beauty
2.	Ratio
3.	Timer
4.	LED Flash button
5.	Camera settings
6.	Tap to view more camera options.
7.	Video tab
8.	Portrait: To take portrait pictures on your device.
9.	Preview thumbnail tab
10.	Camera button
11.	Front-facing/rear-facing camera switch.

To quick launch camera:

1. Swipe down from the top of the screen and select Settings icon
.
2. Scroll and tap **Advanced features.**
3. Tap **Side key** and tap **Quick launch camera**.

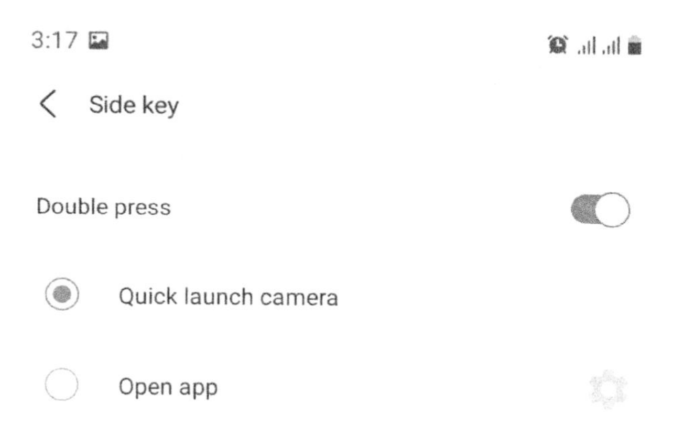

This allows you to launch the camera from any screen by double tapping the power key.

If you tap on the **open app** as shown here, you will be able to customize your side key.

1. Tap on **open app**
2. Select the app

Next time you want to open the selected app, you can double press the side key and it will launch the app. A form of shortcut.

How to record a video

1. When the phone is locked, double-press the **Power Key** to launch the camera app. This is after you've enabled the quick launch camera.

2. Swipe to the video tab then pap on video button to start recording.

3. To zoom in while recording, place two fingers on the screen and spread them apart. To zoom out, move the two fingers closer together.

4. To stop the video, tap the **video button** again.

6. To view your recorded videos, go to **Gallery/Photos** app.

How to take screenshot on your device

This allows you to take a screenshot of your device screen. It allows you to capture everything on your display screen.

There are two ways to screenshot. These are:

- Press and hold the Volume down key and the Power key simultaneously until you see it captured the screen. You can edit the screenshot as the option would be displayed immediately after the screenshot.

You can also share this screenshot to the available apps right from here.

Advanced Features

Advanced features is a section on your device that allows you to have access to some really good features on your Samsung Galaxy A series. To go to **Advanced features**

- Go to **Settings**
- Scroll and tap on **Advanced features**

As shown below

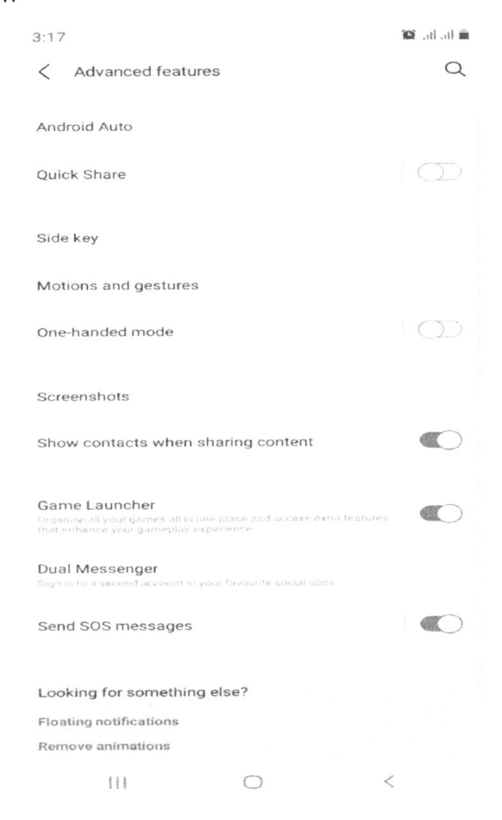

Here you will be able to enable and tweak couple of interesting features.

Tips and Tricks(bonus)

App Icon Badges

Icon badges are the dot that appears on the corresponding app icon whenever there is a notification for an app. We have discussed this trick briefly in this guide earlier.

You can actually customize the app icon badges.

To customize the App icon badges:

- Swipe down from the top of the screen and select Settings icon .
- Tap on **Notification**.
- Tap on **App icon badges** to enable.
- Select the badge style you want

Under the **Notification** you will some other options. You can enable them if interested. Such as **Show snooze option**, and **Suggest actions and replies**. It is advisable to enable these two feature. They are great features to help you enjoy your Samsung Galaxy A12 more. Also you can choose apps you want to be getting notifications from, simply toggle to enable the notification.

Assistant menu

This shows a menu of easy-to-reach buttons that let you replace hardkeys, gestures, and other interactions.

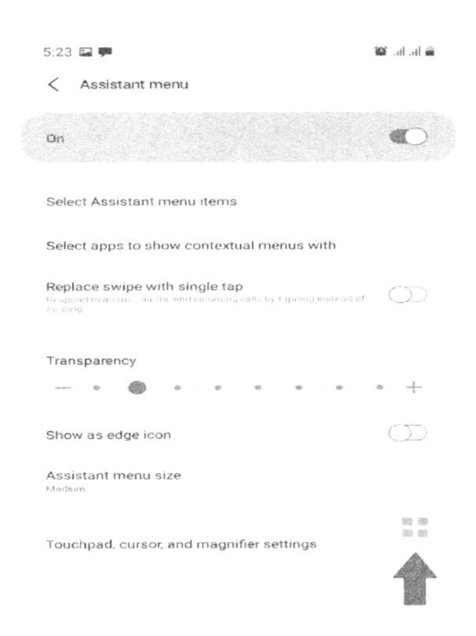

To enable this:

- Swipe down from the top of the screen and select Settings icon .
- Scroll down, tap on **Accessibility**.
- Tap on **Interaction and dexterity**.
- Toggle to enable **Assistant menu**.

Note: You can adjust the **assistant menu** transparency when you tap on the + or – to reduce the transparency. And you can simply drag to adjust.

Interaction control

This feature keeps the focus on a single app. You can block activity from other apps, including calls and notifications. Also you can block touches from a selected area of the screen.

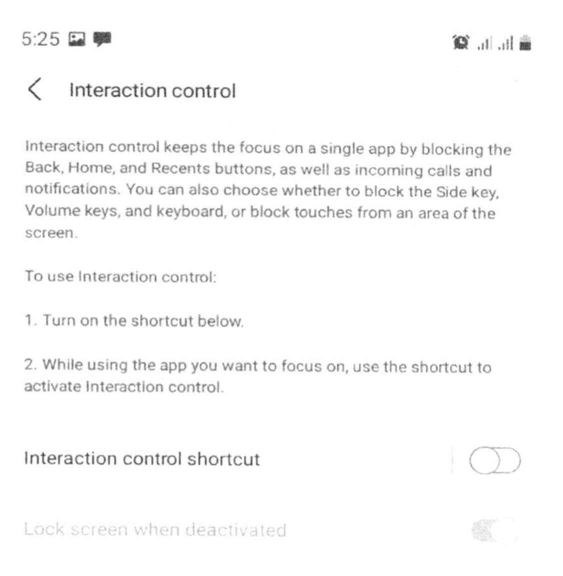

To get this:

- Swipe down from the top of the screen and select Settings icon  .

- Scroll down, tap on **Accessibility**.
- Tap on **Interaction and dexterity**.
- Tap on **Interaction control** then toggle to enable this feature.

Note: To activate or deactivate this feature, press and hold the side and volume up keys for 2 seconds.

Flash notification

This feature allows the camera light or the screen when you receive notifications or when alarms sound.

To get this:

- Swipe down from the top of the screen and select Settings icon ⚙.
- Scroll down, tap on **Accessibility**.
- Tap on **Advanced settings**.
- Tap **on Flash notification**. Toggle to enable the options.

Hide notifications

This feature allows you to choose whether to hide or show notifications, icons, and badges.

To get this:

- Swipe down from the top of the screen and select Settings icon ⚙.
- Tap on **Notifications**.
- Tap on **Do not disturb**.
- Tap on **Hide notifications**, and tap to enable **Hide from notification list**.

Dark mode

To switch to dark mode:

- Swipe down from the top of the screen and select Settings icon ⚙.
- Tap on **Display**.
- Tap on the **Dark** to enable dark mode. As shown below
- Tap on **Dark mode settings** to go its' settings. You can schedule the time you want your device dark mode to be activated.

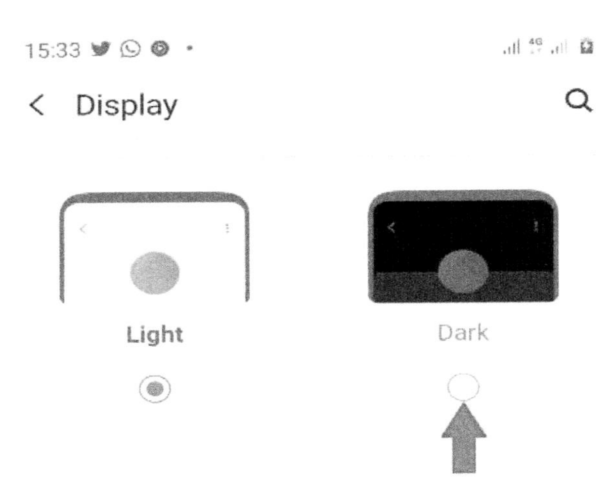
15:33
Display
Light
Dark

One-handed mode

This feature temporarily scale down the display size for easier control of your phone with just one hand.

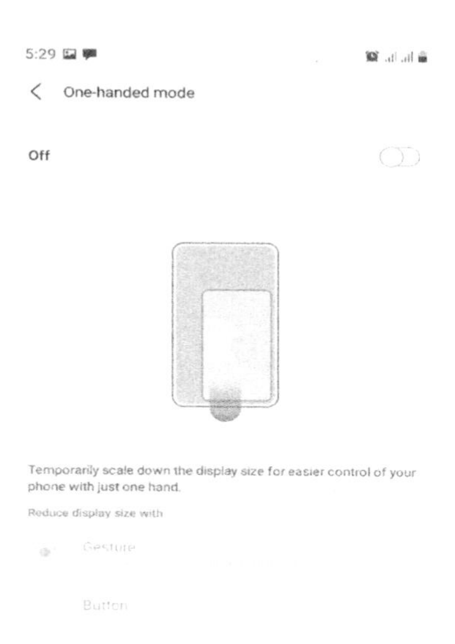

To get this:

- Swipe down from the top of the screen and select Settings icon .
- Scroll down and tap on **Advanced features**.
- Scroll and tap on **One-handed mode**.
- Togle to enable this feature.

Note: You can double tap the home button to activate/deactivate the use of one-handed mode.

Navigation bar

This allows you to use to manage Home, Back, and Recent buttons or use gestures for more screen space.

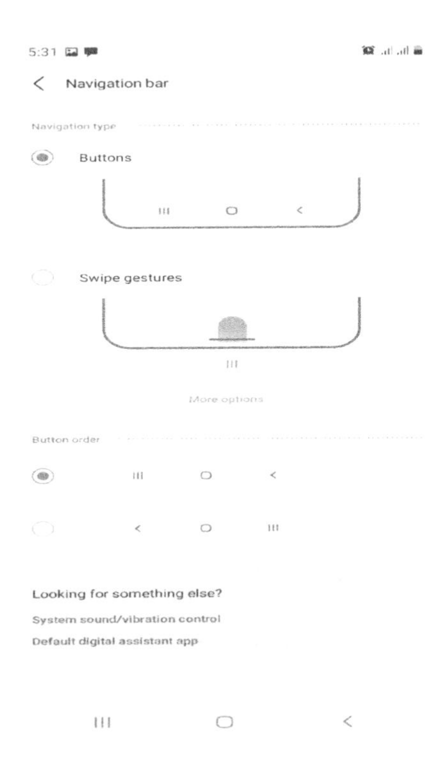

To get this:

- Scroll down and tap on **Navigation bar**.
- Tap to select your choice from the available options.

Change your device name

To change your device name:

- Swipe down from the top of the screen and select Settings icon  .
- Scroll down, tap on **About phone**.
- Tap on **Edit** located behind under the default device name.
- Tap **Done** when you are done typing the name.

Thanks for Reading

Although I have put in efforts to write this guide, and tips and tricks on the Samsung Galaxy A12.
I have no doubt that I have not written everything available on the Samsung Galaxy A12.
However, I will appreciate if you can send me an email on sodiqtade@gmail.com anytime you are unable to perform some task written in or outside this guide concerning the Samsung Galaxy A12.
I will try as much as possible to reply you as soon as I can.

Thanks for reading!

Printed in Great Britain
by Amazon